The Pi

The Pilot's R/T Handbook

How to get your pilot's R/T licence

Chris O'Malley

Airlife
England

First published in the UK in 1991
by Airlife Publishing Ltd.

British Library Cataloguing in Publication Data available
O'Malley, Chris
The pilot's R/T handbook: how to get your pilot's R/T licence.
1. Flying (Aircraft)
I. Title
629.13252

ISBN 1 85310 223 7

Printed in England by Livesey Ltd., Shrewsbury.

Airlife Publishing Ltd.

101 Longden Road, Shrewsbury, England.

Contents

Section A

Background Knowledge, Equipment and Its Use

Section B

LESSONS 1 — 11
Aerodrome Traffic Zones to LARS

Section C

LESSONS 12 — 16
Emergencies, Airways Crossing

Section D

LESSONS 17 & 18
Airways Procedures

NOTE: Airways radio procedures will not be tested for the FRTO licence. This section is for the guidance of those FRTO holders who wish to fly on airways. Assessment will be made during the initial instrument flight test.

Section A

Background Knowledge, Equipment and Its Use

Flight Radiotelephony Operator's Licence (FRTO):
. . "to operate . . . radio telephony apparatus on board any aircraft registered in the United Kingdom . . . VHF only."

Whether you're a private pilot flying only five hours a year or a commercial pilot flying nine hundred, in order to operate from a licensed airfield you must have gained an FRTO licence. And whether it's a Cessna 152 or Boeing 747 that you want to fly, amateur or professional, you still need to reach the same level of proficiency to get your licence.

This manual, drawn from many hours of RTF instruction and practical use, will help you to reach that level of proficiency: and not merely to pass the necessary tests, but to achieve and maintain a level of skill you can take pride in.

LICENCE REQUIREMENTS

To obtain an FRTO licence, if you are a student pilot, you must:

Either as part of your flight training, be instructed in and demonstrate the required level of proficiency in those areas of the RTF (radiotelephony) syllabus listed on page 14. An RTF training record of areas covered will be maintained by the Flight Training Organisation and a final assessment of your ability will be made during your General Flight Test for the Private Pilot's Licence.

Or if you are a student pilot or already hold a PPL, you will need to complete a course of GROUND instruction in all areas of the syllabus, demonstrate your ability during a simulated flight test route with a CAA-approved RTF examiner, and satisfactorily complete a short, multiple-choice/written test. (See example questions on page 58.)

Airways radio procedures will not be tested for the FRTO. Those wishing to fly on airways will be assessed during the initial instrument rating flight test.

HOW TO USE THIS BOOK

This series of lessons has been written to reflect true-to-life practice rather than academic dogma and so will prepare you for the vagaries of everyday radio traffic whilst fostering good habits. It has been written both for self-study and as an instructor's manual.

If you are using this book as a student:

First study Section A on Background Knowledge, Equipment and its Use before starting practical study of Lesson 1 with an instructor. A self-study period should then take place to review that lesson's content, followed in turn by a preview of the lesson to come. Finally, the references to the UKAIP, CAP413 or Aeronautical Information Circulars (AICs) should be consulted.

If you are using this book as an instructor:

Take the student through each lesson after allowing time for self-study, then follow the lesson suggestions listed at the bottom of page 17. Guide your student through the references to the UKAIP, the AICs listed, and any other references made.

THE RADIO INSTALLATION

A typical R/T simulator, such as is found in most flight schools, consists of two panels, each with a volume control, frequency selector, a frequency-in-use display and a press-to-transmit button. In addition, the instructor's panel has a noise-generator control to simulate radio interference, and switches to make an alarm sound or show on the student's panel. This is used to alert the student to a simulated emergency which would then require the appropriate radio message to be sent in response. There is also a provision for a tape player to play pre-recorded radio traffic through the headsets. The student's panel is often isolated in a booth.

Shown opposite is an example of the type of avionics equipment to be found in a modern light training aircraft. As far as R/T is concerned, we're only interested in the communications (COM) equipment outlined, the basis of which is a 720 channel Very High Frequency (VHF) radio — in this case the King KX series. Aeronautical VHF communication is made between the frequencies 118·0 and 135·95mHz (a megaHertz is one million cycles per second). Each available frequency or channel is now separated by 25kHz (25,000 cycles/sec). This spacing allows for 720 individual channels between the limiting frequencies. Whilst only a handful of these might get used in the course of a year in a particular aircraft, having a radio capable of fewer frequencies could severely limit the aircraft's usefulness by preventing its acceptance into certain controlled airspace. To select a frequency, say the 123·425mHz of "Fairoaks Radio" near Heathrow, the outer selector knob is rotated (clockwise to increase), to alter the mHz display to 123. The inner knob is then rotated to change the first two decimal places to ·40 (now 123·40mHz), and can then be pulled out to select a further ·025mHz (123·425). When a radio has a 'FLIP-FLOP" facility such as in the example shown, the new frequency is entered beforehand in the STANDBY (STBY) window on the right-hand side. When that frequency is required, a push on the flip-flop button will swap the STBY and the FREQUENCY-IN-USE over. That way, the same proven radio is used and the old frequency isn't lost from sight in case you need to return to it quickly.

Most radios have a "SQUELCH" control to adjust the sensitivity of the receiver. This is used whenever reception is good to reduce unwanted background noise. By adjusting the knob until the hiss just disappears, the pilot will ensure that speech reaches his ears with maximum clarity. On the other hand, should a received signal be weak, by decreasing squelch the pilot can often make sense of the message despite the increased "static".

Should there have been no radio traffic for him to monitor, the pilot can verify that his receiver is functioning properly by momentarily decreasing squelch. If his radio is receiving the background noise, then it should also receive speech. In the case of modern radios such as the KX series, this function is performed by using the "pull-to-test" control, as squelch is modulated automatically.

THE RADIO PANEL

1. Select the transmitter that you want to use on the rotary switch, seen here on the right-hand side of the AUDIO PANEL pointing to (transmitter) COM 1.

2. Choose whether to hear via speaker or headphones by depressing the upper or lower push-button for whichever transmitter is in use — COM 1 in our example.

3. Switch box 1 ON using the ON/OFF/VOL switch and adjust the volume to a comfortable level. You can use the volume of "squelch" for this if no-one is actually speaking over the radio at the time. Check squelch is apparent on some models by pulling out VOL switch for a few seconds.

4. Use the concentric rotary knobs to "dial" the desired frequency into the STANDBY display and transfer it to the IN USE display with the flip-flop button. Then dial the next frequency to be used into the STANDBY so that it's ready and waiting.

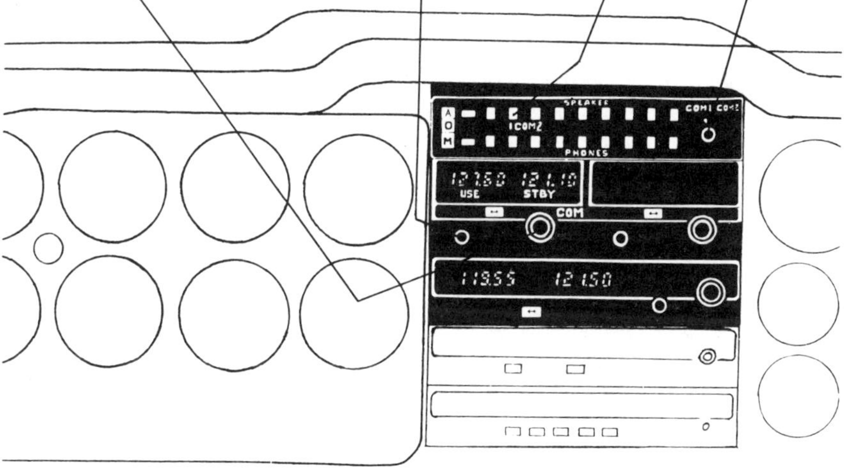

Now you are ready to set up box 2 ready for testing both boxes on two frequencies each. You may also need to check that the INTERCOM is switched to ON, and that both volume and squelch are set.

Most aircraft radio installations offer the option of using either a hand microphone/roof speaker arrangement or of using a headset, which is a combination of earphones and boom mike, with a press-to-transmit button on the steering yoke. This is the most convenient system to use as it leaves the hands free to note down information and elminates the likelihood of dropping the hand mike at an inconvenient moment, such as on "finals".

Choice of radio or system to use is made on an AUDIO PANEL, such as the one shown on the previous page. Rotating the right-hand knob to COM 1 or 2 selects the transmitter, either the top radio in the stack which is by custom box 1, or box 2. Choice of speaker or headphones is made by depressing the COM button for that radio, on the audio panel. Volume and squelch are set on the radio as described earlier.

There are weak links in each system, such as at the points at which the cables emerge from the jackplug bodies. These are subject to considerable tugging and flexing, and connections sometimes fracture. Should such a fault occur, necessitating a change over to the alternate system, it will, of course, be necessary to switch systems on the panel accordingly.

When we speak, it is useful to hear our own voices so that we can modulate volume, expression and pronunciation. Normally, this would be impossible due to the engine noise or the headphones that we wear. However, radio manufacturers have built in a facility called SIDETONE to allow this. Hearing sidetone during transmissions whilst using headphones is an indication that the transmitter is functioning correctly. Should sidetone cease, then transmitter failure should be immediately suspected. (See Radio Failure p.51.)

Another useful gadget often found on the instrument panel is the EMERGENCY-PRESS-TO-TRANSMIT button or EPTT. This is used to transmit by when wearing headsets, should the button on the yoke prove faulty.

You should also come across the INTERCOM, which enables aircrew to converse easily with one another when using headsets. Some intercoms allow the back seat passengers to listen in as well. The ON/OFF switch for both functions can usually be found on the audio panel or on the right-hand side of the panel on the front of the intercom unit itself.

As you may know, delicate avionics equipment can be damaged by surges of electrical current during engine start. By ensuring that all avionics are switched off before starting, such damage will be avoided. If fitted, this is best done by the use of an AVIONICS MASTER SWITCH. This allows all of the COM and NAV equipment to be switched on or off simultaneously and yet left at their correct volume settings. Avionics are at their most trouble-free when used on every flight, but sulk if they are left out of anything. The avionics master switch helps you to switch everything on for every flight and so avoids such anti-social behaviour.

Remember that VHF transmissions (meaning those between 30 and 300mHz) are approximately line-of-sight only; and that their reception can be distorted or blocked by the curvature of the earth or by high terrain unless the aircraft has sufficient altitude.

Unfortunately, there are only so many frequencies available between 118·0 and 135·95mHz, even with the 25kHz spacing of the 720 channel system. This means that many airfields have common frequencies, but to provide protection from causing interference to one another, they are only allocated to airfields a considerable distance apart. Their "protected range" is intended to prevent interference from an aircraft at 4000′ and 25 miles away on the same frequency. This system is generally successful but there are airfields at which you might hear unconnected radio traffic from an aerodrome many miles away, such as used to happen at Shipdham* in Norfolk, and Fenland in Lincolnshire. Both shared the frequency of 123·05. At such airfields it is especially important to state the station you are calling in every call.

* Shipdham frequency changed 1990 to 119·55 because of this problem.

Further Study

UKAIP Vol 1, RAC 6-1 and Schedule 5 of the ANO which show the equipment required for flight in different circumstances whilst in controlled airspace.

Note that the demand for aeronautical VHF channels will mean further "channel-splitting" from 25kHz spacing to 12·5kHz spacing around 1998.

USING THE RADIO

There is a delightful scene in the old Ealing comedy *The Lavender Hill Mob* where Stanley Holloway replies to a message with the word "Edgar". He meant, of course, "Roger", which is simply radio shorthand for "I have received all of your last transmission". This is standard ICAO phraseology and each has a specific meaning. Here is some more of the same. Learn them.

ACKNOWLEDGE	Let me know that you have received and understood this message.
AFFIRM	Yes.
APPROVED	Permission for proposed action granted.
BREAK	Indicates the separation between messages.
CANCEL	Annul the previously transmitted clearance.
CHECK	Examine a system or procedure (no answer is normally expected).
CLEARED	Authorised to proceed under the conditions specified.
CONFIRM	Have I correctly received the following . . . or Did you correctly receive this message?
CONTACT	Establish radio contact with . . .
CORRECT	That is correct.
CORRECTION	An error has been made in this transmission (or message indicated). The correct version is . . .
DISREGARD	Consider that transmission as not sent.
HOW DO YOU READ	What is the readability of my transmission?
I SAY AGAIN	I repeat for clarity or emphasis.
MONITOR	Listen out on (frequency).
NEGATIVE	No or Permission not granted or That is not correct.
OVER	My transmission is ended and I expect a response from you.
OUT	This exchange of transmissions is ended and no response is expected.

PASS YOUR MESSAGE	Proceed with your message.
READ BACK	Repeat all, or the specified part, of this message back to me exactly as received.
REPORT	Pass requested information.
REQUEST	I should like to know . . . or I wish to obtain . . .
ROGER	I have received all your last transmission. NOTE: Under no circumstances to be used in reply to a question requiring a direct answer of AFFIRM or NEGATIVE.
SAY AGAIN	Repeat all, or the following part of your last transmission.
SPEAK SLOWER	Self explanatory.
STANDBY	Wait and I will call you. NOTE: No onward clearance to be assumed.
VERIFY	Check and confirm.
WILCO	I understand your message and will comply with it.

This list has been carefully chosen to avoid confusion with similar-sounding words since much of the information relayed by R/T is of too great an importance to leave to assumption and guesswork. And should confusion nonetheless arise, due for example to poor reception, it may even become necessary to spell a word out. To prevent misunderstanding when spelling, each letter of the alphabet is nomenclated: Alpha, Bravo, Charlie and so on. The letters B and P, for example, can be difficult to distinguish between by ear, but change them to Bravo and Papa and they are easily identified.

Learn the following until they are second nature:

Letter	*Word*	*Pronounced as*
A	Alpha	AL-fah
B	Bravo	BRAH-VO
C	Charlie	CHAR-lee
D	Delta	DELL-tah
E	Echo	ECK-oh
F	Foxtrot	FOKS-trot
G	Golf	GOLF
H	Hotel	ho-TELL
I	India	IN-dee-ah
J	Juliet	JEW-lee-ETT
K	Kilo	KEY-lo
L	Lima	LEE-mah
M	Mike	MIKE
N	November	no-VEM-ber
O	Oscar	OSS-cah
P	Papa	Pah-PAH
Q	Quebec	keh-BECK
R	Romeo	ROW-me-oh
S	Sierra	see-AIR-rah
T	Tango	TANG-go

U	Uniform	YOU-nee-form
V	Victor	VIK-tah
W	Whisky	WISS-key
X	X-ray	ECKS-RAY
Y	Yankee	YANG-key
Z	Zulu	ZOO-loo

For the same reason, numbers are pronounced in a particular manner:

Number	*Pronounced as*
0	ZEE-ROW
1	WUN
2	TOO
3	TREE
4	FOE-er
5	FIFE
6	SIX
7	SEV-en
8	AIT
9	NINE-er
Decimal	DAYSEEMAL

TIME is given as minutes past the current hour, UCT (Universal Co-ordinated Time, see also GMT and "Zulu"). For example, an estimated-time-of-arrival (ETA) fifteen minutes after 0800 would be given as "Time Wun Fife". If the ETA was to be an hour further on it would be spoken in four figures, "Zeerow Niner Wun Fife".

In fact, the clearest way to communicate any number of two or more figures is to say them separately, e.g. a QFE of 992 would be "Niner Niner Too" or a height of 1580′ spoken as "Wun Fife Ait Zeerow". On the other hand, whole thousands can be given as "Touzand", hence "Wun Too Touzand" for 12,000. This also minimises your transmission time.

Further Study

UKAIP Vol. 2 COM 4-1.

THE RTF TRAINING SYLLABUS REQUIRED BY THE CAA

		See Page
Pre-Flight		
1.	Use of the UK AIP COM section and familiarisation with the appropriate parts of the RAC section	11 *et al*
2.	Familiarisation with aircraft radio equipment	8
3.	Microphone technique and listening-out	10
4.	Call-signs and abbreviations	16
5.	Phonetic alphabet and standard words and phrases	11, 12
Departure Procedures		
6.	Radio checks	17, 72, 73
7.	Taxy instructions/information and read back	16, 17
8.	Holding	17, 21
9.	Departure clearance	17, 25
En Route Procedures		
10.	Frequency changing	17, 41
11.	Height reporting	18, 71
12.	Position reporting	18, 71
13.	Use of FIS	31, 62
14.	Use of LARS	38
15.	Use of SSR	30, 39, 44, 48, 55, 71
	MATZ penetration	30
	SVFR flight	28
	SRZ entry	26
16.	VDF (QDM, QTE)	32
Circuit Procedures		
17.	Joining circuit	19, 21
18.	Landing clearance	19, 23
19.	Orbit, Extend	21, 23
	Touch-and-go	21
	Go around	22, 23
20.	Vacating runway	23
Emergency Procedures		
21.	Uncertainty of position	49
22.	Radio failure	51
23.	Practice Pan	48
24.	Mayday (Ground practice only)	41

Further Study

AIC 44/1986 (White 199) 5 June.

Section B

Lessons 1 – 11
AERODROME TRAFFIC ZONES TO LARS

Lesson 1: Leaving an aerodrome traffic zone
Readability scale, taxi clearance,
airfield data, readback list, radio check

Lesson 2: Entering an aerodrome traffic zone
PHACER call

Lesson 3: RT in the circuit
Circuit diagram, circuit calls, variations

Lesson 4: Exiting a control zone
Non-standard R/T, reporting points

Lesson 5: Entering a control zone
Ground control

Lesson 6: Special VFR clearances
SVFR requirements, entry points

Lesson 7: MATZ penetration
Zone dimensions, clutch QFE,
identification

Lesson 8: VDF
QDM, QDR, QUJ, QTE

Lesson 9: ATIS
ATIS Alphabet
Inbound and outbound

Lesson 10: VOLMET
Regional frequencies and times

Lesson 11: LARS
Limitations,
service offered,
"handing over"

Lesson 1: Leaving an Aerodrome Traffic Zone

Most instructors introduce students to the radio as soon as possible so your first experience of aeronautical radiotelephony will probably come during your first few flying lessons.

Opposite is such an exchange. It begins at (1) with your INITIAL CONTACT CALL to the air traffic controller (ATC) in the tower, alerting him to your presence on the field, and testing, or "pre-flighting" your transmitter at the same time. His reply (2) will vary according to the scale below:

Readability 1 . . . . . Unreadable
Readability 2 . . . . . Readable now and then
Readability 3 . . . . . Readable but with difficulty
Readability 4 . . . . . Readable
Readability 5 . . . . . Perfectly readable

At (3) you inform the controller how his transmission is being received by you and you ask for TAXI CLEARANCE and AIRFIELD DATA (runway in use, pressure settings). The clearance and the settings are of importance to safety and so are read back to the controller so that he can check that they have been received correctly, as are all the items on the following list:

Level Instructions
Heading Instructions
Speed Instructions
Airways or Route Clearances
Clearance to Enter, Land On, Take-Off on
Backtrack or Cross an Active Runway
SSR Operating Instructions
VDF Information and Frequency Changes. If you don't read them back you will be asked to do so.

Note

A. At a large airfield your initial call as in (1) might also include your whereabouts on the field. For example,
"Biggin Hill Tower, G-ABCD on the Main Apron.
Request Radio Check box 1 on 134·8 . . ."

Furthermore, on an even bigger airfield, you may have to listen to an ATIS or have to contact a GROUND frequency first (see ATIS p.34). See also Glossary in appendix for any words you might not understand.

B. ATC personnel may shorten a callsign as in G-CD. If they do this then you may use the abbreviated version also. However, should there be two aircraft on frequency with callsigns that have similar endings, then the controller will insist on the use of the full callsign to avoid ambiguity. You can imagine the confusion that might arise if the two aircraft callsigns, G-ABCD and G-BBCD were shortened when both were in the circuit.

C. To simplify tuition in this manual, all personnel operating ground radio stations will be referred to as AIR TRAFFIC CONTROL or ATC. Their differing roles and responsibilities will be explained later.

D. Should your transmission be received at less than strength 5, your instructor may want to make a test transmission from his side, or even try another headset or microphone altogether. Be aware however, that your transmission may have been blocked by large buildings such as hangars.

E. You should always state your frequency in your initial call so that ATC, who may be monitoring both a TOWER and an APPROACH frequency, will know which one to use. And if you have the wrong frequency "dialled in", anyone listening can tell you of your error right away.

LEAVING AN AERODROME TRAFFIC ZONE	
YOU – frequency 118·9	**AIR TRAFFIC CONTROL**
(1) Norwich Tower, G-ABCD Request RADIO CHECK on 118·9, over.	(2) G-ABCD, Norwich Tower. Reading you fives.
(3) G-ABCD readability five also. Request AIRFIELD DATA and TAXI CLEARANCE for VFR flight to the south-east.	G-CD, roger. Taxi to hold short of Runway 27, QNH1030.
"Taxi, hold short of 27, QNH1030," G-CD.	

G-CD is ready for departure.	G-CD is cleared for take-off. Left turn-out, VFR. Wind 270/10kts.
"Cleared for take-off. Left turn out, VFR," G-CD.	

	G-CD was airborne 35. Change now to Norwich Approach on 119·3.
"Changing to 119·3," G-CD.	

(Change frequency to 119·3 and pause before speaking to ensure that you don't transmit over another pilot's message).

Norwich Approach, this is G-ABCD on 119·3.	G-CD, roger. Report when you wish to re-enter.
Wilco, G-CD out.*	

(* The use of the words "over" and "out" is discretionary and is left up to the individual pilot or operator.)

Instructor

Go through the above text using both sides of the page. When your student is familiar with the layout and what he/she is expected to do:

i) cover the left-hand side and go through it again.

ii) when ready, cover both sides and go through it once more, substituting new information for the airfield data and taxi instructions, e.g.

> "G-CD, roger. After the Aztec passing you from left to right, taxi to HOLD SHORT of Runway 09, QNH1022."

Only when you're satisfied that the student understands this lesson, go on to Lesson 2.

Further Study

Study pp33-42 CAP 413 in preparation for next lesson.

Lesson 2: Entering an Aerodrome Traffic Zone

On your return to the aerodrome from the training area, ready to join the traffic and to land, the controller will need to know where you are, what height you are at and your ESTIMATED TIME OF ARRIVAL (ETA). In returning this information you will find the use of a mnemonic invaluable. There are no set rules about the use of these aides-memoire, you can even make up your own, but one which I would recommend is the widely-used "PHACER".

P — Position
H — Heading
A — Altitude
C — Conditions
E — Estimate
R — Request

You would probably have been listening out on frequency so it will just be necessary to make a brief Contact Call as in (a). When the controller replies by asking you for your details you can then make your PHACER call, prefacing it with your callsign and aircraft type.

When you make your initial call, be prepared for the controller to ask you to "STANDBY". This simply means that he is occupied in another task for the moment (he may be talking to other aircraft on another frequency), but that he will return to you very soon.

Instructor
First, explain the relevance of each element of PHACER. Then, as in Lesson 1, go through the text with your student until it is fully understood. Then go through it again, covering each side in turn, and substituting other information, e.g. ask your student to suggest likely information for his PHACER call, then reply to his transmission with "G-DE, roger. Call crossing the city boundary. Expect to join right-hand, downwind for runway 09, QFE992" . . . and so on.

Further Study
Read pp 72-73 CAP413.

ENTERING AN AERODROME TRAFFIC ZONE

YOU – frequency 119·3	**AIR TRAFFIC CONTROL**
(a) Norwich Approach, this is G-BCDE, over.	G-BCDE, Norwich Approach. Pass your message.
G-BCDE is a Cessna 152 P – Position: Overhead Cantley H – Heading: 330 Magnetic A – Altitude/FL: 1800′ C – Conditions: VMC E – Estimate: Estimating field 45 R – Request: Request re-joining instructions.	G-DE, roger. Call FIELD IN SIGHT. Join LEFT BASE for runway 27, QFE 1026, over.
Wilco, "left base for Runway 27, QFE 1026," G-DE.	

G-DE has the field in sight, over.	G-DE, roger. You are cleared to FINAL number 1. Call before turning final.
"Cleared to final, number 1. Call before turning final," G-DE.	

G-DE is ready to turn final.	G-DE, you are cleared to land, wind 260/15.
"Cleared to land," G-DE.	
	G-DE landed 25. Take the first turn on your left.
Wilco. May we have some fuel, please? G-DE.	G-DE, roger. I'll arrange that for you now.

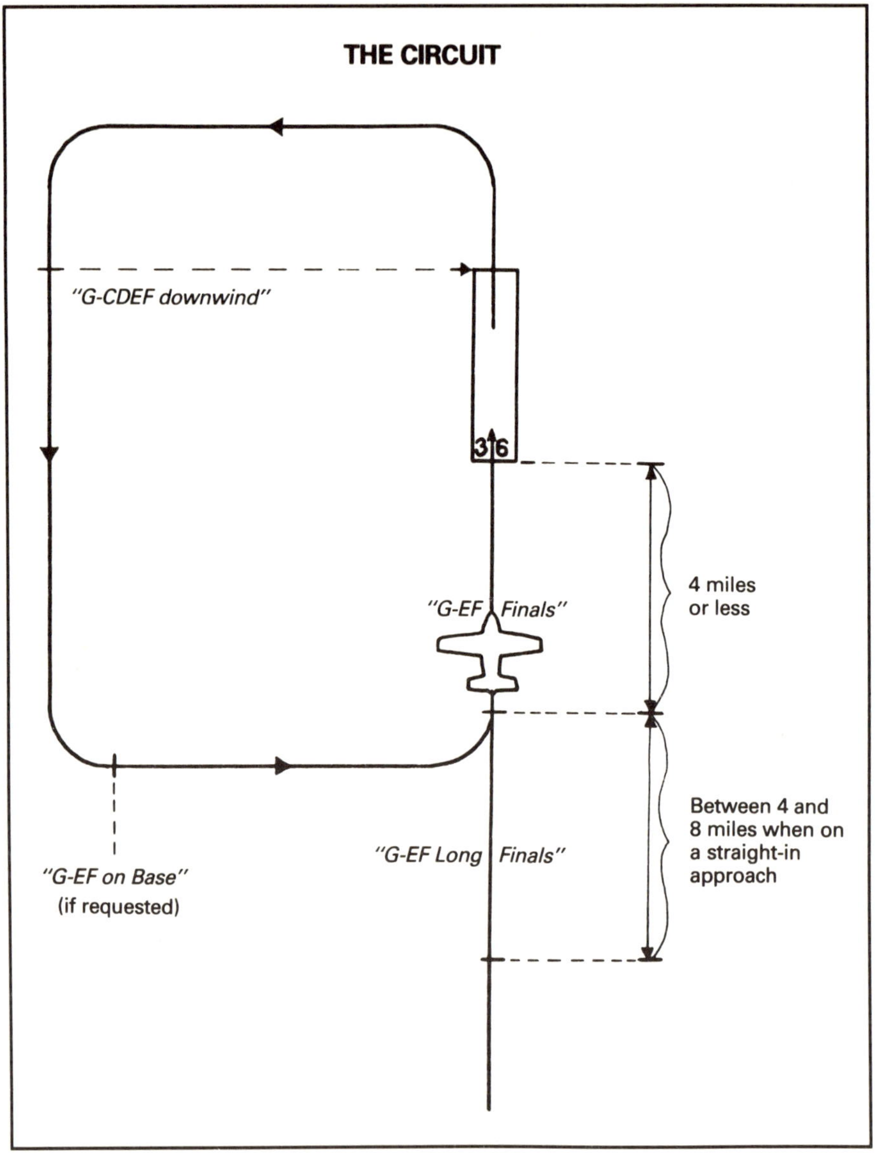
THE CIRCUIT
"G-CDEF downwind"
36
"G-EF Finals"
4 miles
or less
"G-EF on Base"
(if requested)
"G-EF Long Finals"
Between 4 and
8 miles when on
a straight-in
approach

Lesson 3: The Circuit

The busy circuit, with its high demands on concentration and its heavy workload, is no place to learn R/T by trial and error. That would mean too much trial for everyone concerned. Common sense dictates that the student pilot be introduced to as many of the variations likely to be encountered as soon as possible for when he flies solo. Forewarned is forearmed.

The exercise begins with the aircraft, G-CDEF, on the ground with the engine running.

(1) Remember that the golden rule of all clearances, whether on the ground or in the air, is that the clearance is to that point stated and no further. What that means in this case is that we are cleared to line up and hold — but not to take off.
(2) Make a note of the circuit direction.
(3) The "DOWNWIND" call is made when abeam the upwind end of the runway in use (see diagram opposite). Should other radio traffic prevent this then the call should be made as soon as possible on the Downwind leg, calling "LATE downwind" if necessary. It helps the controller if we can state our intention here, i.e. whether "to land" or to "touch-and-go" or "low approach and go-around". Use the standard terminology though and not the "smash-and-go" that I heard one disheartened student use!
(4) Having been cleared to the final approach, the "finals" call is made when the turn is completed.
(5) Remember the "readbacks" listed in Lesson 1.
(6) This call is made only after the overshoot is safely initiated.

Further Study
Read pp 43-50 CAP413.

Instructor
This exercise is best taught in a series of simulated "circuits". Once the student understands the basic exercise then you can introduce the many variations which may be encountered in the circuit. Run through several circuits one after another, altering the air traffic instructions on each one to be as realistic as possible.

LESSON 3: THE CIRCUIT	
YOU	**AIR TRAFFIC CONTROL**
Norwich Tower, G-CDEF RADIO CHECK on 118·9, over.	G-CDEF, Norwich Tower reading you fives.
G-CDEF reading you fives also. Request AIRFIELD DATA and TAXI CLEARANCE for the circuit, over.	G-EF, roger. Taxi to holding point for Runway 27, QFE1011.
Taxi to holding point for 27, QFE1011, G-EF.	G-EF, that is correct.

G-EF is ready for departure.	G-EF, roger. Line up and hold.
(1) "Line up and hold," G-EF.	

(2)	G-EF is cleared for take-off. Left-hand turn out, wind 260/10kts.
"Cleared for take-off, left turn out," G-EF.	G-EF, that is correct.

(3) G-EF is downwind for touch-and-go.	Roger. G-EF is cleared to final, number one. Call finals.
"Cleared to finals number one, call finals," G-EF.	

(4) G-EF, finals for 27.	(5) G-EF is cleared to touch-and-go, runway 27, wind 270/15kts.
"Cleared to touch-and-go," G-EF.	

(6) G-EF going around.	Roger, G-EF.

Example 1

STUDENT	INSTRUCTOR
(2)	G-EF, roger. Are you ready for an immediate (departure) on runway 22?
Affirm, G-EF.	Roger. G-EF is cleared for an immediate take-off, runway 22, right-hand turn-out, wind 270 less than 5kts. The landing traffic is a 757 on a six-mile final.
(Student acknowledges).	

Instructor: Discuss the implications of an immediate take-off.

Example 2

(4)	G-EF is cleared to final, number three. Number one is a Pa32 on short finals, number two is a Jodel joining Right Base ahead of you.
"Cleared to final, number 3." I have the other traffic in sight, G-EF.	G-EF make one left-hand orbit to allow traffic to join Right Base.
Making one left-hand orbit, G-EF.	Thank you, G-EF. When your orbit is complete, join Right Base for runway 04. Call before turning finals.
(Student acknowledges).	

Example 3

(5)	G-EF GO AROUND. I say again, go around due to unauthorised crossing traffic. Acknowledge.
(Student acknowledges).	

Example 4

(On this circuit, the student has indicated his intention to land, i.e. to make a FULL STOP and has made his "finals" call).	
(4)	G-EF, you are cleared to land, wind 230/10kts.
"Cleared to land," G-EF.	

	G-EF landed 05. You are cleared to backtrack to the Western Taxiway. Call when clear.
"Cleared to backtrack", G-EF.	

G-EF has vacated runway 04.	(ATC acknowledges).

Discuss the implications of how and when to call runway vacated, e.g. at night or in poor visibility or when hidden from tower.

Finally, the student should be forewarned that it is possible, though a rare occurrence, for ATC to order that all traffic must clear the circuit in the event of an emergency, or that in such an event it may be necessary for circuit traffic to change to another frequency.

Lesson 4: Exiting a Special Rules Zones

This exercise takes place at a busier aerodrome than seen so far — one with its own CONTROLLED airspace due to the volume of its public transport traffic, and subject to Instrument Flight Rules.

VFR traffic leaving a special rules zone will be given a clearance to do so providing that the departure can be made within specified limits. The pilot will be cleared to leave VFR, via a known VFR reporting point such as a lake or a motorway intersection for incoming IFR traffic.

The lesson begins with a radio check on the tower frequency (1). In this exchange, the pilot has included non-standard R/T terminology, and so has the controller (2). Now, we are brought up to understand that such minor pleasantries as this are only polite and help to make the world go around. It certainly appears harmless enough. However, the custom of exchanging small talk is not suited to radiotelephony as idle chatter delays the next person from getting his call in. If you're making an instrument approach in poor conditions or are on short finals and waiting to receive a landing clearance, it can be quite frustrating and even dangerous to have to listen to someone indulging in small talk. It sounds pedantic I know, and a bit spoilsport-ish, but don't pick up the habit in the first place. Keep superfluous chat to the minimum.

Our VFR clearance comes in (3), which we, of course acknowledge — as we do with the instruction to change frequency (4).

Once clear of the zone we are passed any further traffic information and we may be advised to change to a third frequency, that of a radar facility as in this example.

Further Study

See also pink AIC 65/1983, "RTF Phraseology and Discipline". (This AIC has now lapsed and not been replaced but is well worth looking for.)

Exiting a Special Rules Zone

YOU	AIR TRAFFIC CONTROL
(1) Cardiff Tower good morning, G-DEFG. Radio check on 125·0, over.	G-FG Cardiff Tower, (2) good morning, reading you fives.
G-FG reading you fives also. Request airfield data and taxi clearance for VFR flight to the north-west, over.	G-FG, roger. Taxi to hold short of runway 30. QFE1001, QNH1008.
Roger, "taxi hold short 30, QFE1001, QNH1008, G-FG.	

G-FG ready for departure.	G-FG, roger. Your clearance will be to leave Cardiff Zone VFR, via St. Hilary mast, not above 1500′.
(3) Roger, "Cleared to leave the zone VFR via St. Hilary mast, not above 1500′," G-FG.	G-FG, that is correct. Line up and hold.
"Line up and hold," F-FG.	

	G-FG, you are cleared to take-off, right turn out, wind 290/10.
"Cleared to take off, right turn out," G-FG.	

(4)	G-FG was airborne 26. Change now to Cardiff Zone on 125·8.
"Changing to 125·8," G-FG.	

(Frequency change and pause).

Cardiff Zone, this is G-DEFG. on 125·8, over.	G-FG, roger. Report St. Hilary mast. One light aircraft to the west at 2000′. Wessex 1002. What is your intended level of operation?
Report St. Hilary mast, traffic copied. Wessex 1002. G-FG will be operating at 3000′.	G-FG, roger.

G-FG is at St. Hilary mast, over.	G-FG, roger. No height restriction. Report when you wish to re-enter.
No height restriction. Wilco, G-FG.	

Lesson 5: Entering a Special Rules Zones

This exercise is a straightforward reversal of the previous procedures in Lesson 4.

It begins with the Initial Contact Call at (1). The pilot passes his details using a PHACER call, is instructed to join via a known point, and is given a height limitation.

(2) is a requested position report.

The instruction to change to Tower (3) is generally given when the pilot has the field in sight. Note that the instruction is to "downwind" and no further. Should he be unable to make his "downwind" call at the point adjacent to the take-off part of the runway (perhaps because of waiting for a gap in RTF traffic), the pilot may make it later as a "G-GH LATE downwind" call. He could also have to extend his downwind leg to accommodate other traffic on final approach.

(5) Clearance to final approach.

(6) Should you be landing at an airfield you are not familiar with or where the taxiways are not obvious, then ask for "progressive taxi instructions". If you try to muddle through then remember, it is quite possible at some airports such as Southend, to end up taxi-ing onto a public road or into a car park!

Further Study

pp45, 49, 50 CAP413.

Entering a Special Rules Zone

YOU	AIR TRAFFIC CONTROL
(1) Cardiff Zone, this is G-EFGH on 125·8, over.	G-EFGH, Cardiff Zone. Pass your message.
G-EFGH is a Robin DR400, (P) overhead Bridgend (H) heading 110 (A) at 2000′ (C) VMC (E) estimating zone boundary at 55 (R) request joining instructions.	G-GH, roger. Join via St. Hilary mast, not above 1500′, for right-hand downwind to runway 30, QFE1002.
Roger, "cleared to join via St. Hilary, not above 1500′ for right-hand downwind to runway 30, QFE1002," G-GH.	G-GH, that is correct. Report St. Hilary.
"Report St. Hilary," G-GH.	
(2) G-GH is at St. Hilary mast, over.	G-GH, roger. Change now to Cardiff Tower on 125·0.
Changing to 125·0, G-GH.	
(Frequency change).	
(3) Cardiff Tower, G-EFGH on 125·0, over.	G-GH, roger. Report downwind right-hand for runway 30, QFE1002.
Wilco, "downwind right-hand for 30, QFE1002," G-GH.	
(4) G-GH is downwind.	G-GH is cleared to final, number 1.
"Cleared to final, number 1," G-GH.	
(5) G-GH is on finals.	G-GH is cleared to land. Wind 280/15.
"Cleared to land," G-GH.	
(6)	G-GH landed 03. Take the first taxiway on your left. Contact GROUND on 121·7.*
"First taxiway on left, and contact Ground on 121·7," G-GH.*	

*(For the purposes of the exercise only. There is at present no Ground Control frequency at Cardiff.)

Lesson 6: Special VFR Clearances (SVFR)

Assume in this exercise that you are the pilot of a Pa28 Warrior flying from Sleap to Manchester International. Not being qualified to fly without restriction in controlled airspace, you must therefore request a SVFR Clearance. This is a concessionary clearance granted to you only when the controller considers that current traffic and weather conditions permit your entry into controlled airspace.

In accepting a SVFR clearance you must comply with ATC instructions. You may be given a height restriction (i.e. not above X number of feet), a reporting point, and will be instructed to remain "clear of cloud and in sight of the ground". This means that you must remain in flight conditions which will enable you to VISUALLY determine your flight path and to avoid obstructions.

When flying on a SVFR clearance within the Controlled Airspace of certain aerodromes you will be directed by radar, such as at Manchester International in the exercise. Remember that a SVFR clearance does not exempt the pilot from his responsibility for ATZ avoidance. He must also obey Rule 5 except when cleared below 1500'.

Remember too, that the aerodrome that you left in uncontrolled airspace at the beginning of your flying lesson may, if it is the venue for a Royal Flight (and so subject to Rule 21), be controlled airspace on your return, necessitating a request from you for a SVFR clearance.

The exercise comprises an Initial Contact Call, a PHACER call, and a readback of ATC instructions.

Instructor

Once the student understands this exercise, adapt it to local landmarks and alter the height restrictions and pressures to give practice in reading back.

You could also introduce radar vectors here and issue directions to keep up speed on base leg or final approach to give the exercise the feel of a large aerodrome.

Further Study

Read p74 CAP413
UKAIP Vol 1, RAC 1-6 and 1-7

SVFR	
The aircraft flown is a Pa28 Warrior and the frequency changed to is that of Manchester Radar on 119·4mHz.	
Manchester Radar, this is G-FGHI.	G-FGHI, Manchester Radar. Pass your details.
G-FGHI is a Pa28 Warrior VFR from Sleap presently overhead Whitchurch Heading 050 3000′ on the Barnsley VMC Estimating Manchester at 45 and requesting SVFR clearance.	Roger, G-FGHI. You are cleared to enter the zone via the Entry Lane at Sandbach. Maintain VFR. Not above 1500′ on the Manchester QNH of 1011. Call overhead Sandbach.
"Cleared to enter the zone via Entry Lane at Sandbach. VFR, not above 1500′ on QNH1011. Will call overhead Sandbach," G-FGHI.	G-FGHI, that is correct.

Manchester Radar, G-FGHI is overhead Sandbach at 1500′.	

Note that in this example, the controller has not abbreviated the callsign of the aircraft and so it must be used in full.

Lesson 7: MATZ Penetration

The MILITARY AERODROME TRAFFIC ZONE (MATZ) is a cylindrical column of airspace of 5nm radius and 3000′ in height. Aligned with the approach to the aerodrome's major runway is a rectangular stub measuring 5nm by 4 and extending from 1000 to 3000′ in height.

Civilian traffic wishing to penetrate a MATZ must call the military aerodrome controller either 15nm or 5 minutes flying time, whichever is greatest, before crossing the MATZ boundary. Pilots entering a MATZ "must comply with any instructions issued by the controller, maintain a listening watch on the frequency allocated until clear of the MATZ" and may be passed a "CLUTCH QFE".

The clutch QFE is a pressure setting common to military aerodromes sited close together. It is the QFE of the highest aerodrome in the combined zone.

Outside of the hours published in the AIP (RAC section) pilots should "proceed with caution" should there be no reply to two consecutive calls.

A. NOTE that the initial contact call and the request for MATZ penetration are made all in one (1).

B. Having received your details and ETA to the Zone Boundary, the controller will wish to determine your position. If not transponder equipped, you may be given headings to steer so that you may be identified on radar. If you are so equipped, then you will be given the clutch QFE and a discrete code to squawk.

Within the zone you may be given headings or height changes to fly in order to avoid fast-moving traffic. When you are clear you will be instructed to resume your own navigation.

Instructor

When the student can complete the exercise without needing to consult the text, then alter the instructions that you give.

In (2), for example, you could change the code to 6361 or you could issue heading instructions for radar identification, having prewarned the student that his aircraft is not transponder equipped. e.g.

"Roger. G-IJ. Turn onto heading of 210 for radar identification."

(Student acknowledges new heading). "G-IJ, further right turn on 240."

(again acknowledged).

"G-IJ, you are identified, resume own navigation, no conflicting traffic."

In (3):

"G-IJ, you are identified. Call Zone Boundary. Maintain 1500′ on clutch QFE of 998 millibars. Turn now onto heading of 200 to avoid fast-moving traffic now in your two o'clock moving right to left."

(Acknowledgement).

"G-IJ, that traffic is now clear of you, resume own navigation."

Further Study

UKAIP Vol 1, RAC 3-2-12 to 3-2-15.

MATZ PENETRATION

The aircraft flown is a Mooney 201. The exercise begins with the Initial Contact Call on 134·3.

YOU	AIR TRAFFIC CONTROL
(1) Valley Approach, this is G-GHIJ. (R) Request MATZ PENETRATION, over.	G-GHIJ Valley Approach. Pass your details.
G-GHIJ is a Mooney 201 VFR Haverfordwest to a private strip at Amlwch. (P) overhead Nefyn (H) heading 150 (A) 2000′ on 995 (C) VMC (E) zone boundary 25.	Roger, G-IJ. Are you transponder equipped?
Affirm, G-IJ.	(2) G-IJ Squawk 6563 for identification.
"Squawk 6563," G-IJ	(3) G-IJ, you are identified. Maintain 2000′ on CLUTCH QFE of 1003. Call zone boundary.
"Maintain 2000′ on QFE1003 and call zone boundary," G-IJ.	

G-IJ is at zone boundary.	G-IJ, roger. Call zone boundary outbound.
"Call zone boundary outbound," G-IJ.	

(4) G-IJ is at the zone boundary outbound.	G-IJ, roger. Resume own navigation. Squawk 4321 and change now to en-route frequency.
Squawking 4321 and changing to London Information on 124·7, G-IJ.	

Lesson 8: VDF

Very High Frequency Directing Finding (VDF) is a tremendously useful aid to navigation. Equipped only with a simple COM set — or even an emergency hand-held transceiver, the pilot can obtain a position fix in poor visibility, find his way to the facility, and even make a simple "cloud-break" approach to a suitably-equipped aerodrome.

The operator in the tower reads the information directly off the cathode-ray tube display which responds to each transmission from the aircraft. A switch, spring-loaded to display the more-commonly-required **QDM** —

> **"A MAGNETIC heading to steer TO the facility in conditions of nil wind" —**

— can be turned to display a **QDR** —

> **"A MAGNETIC bearing FROM the facility to the aircraft" —**

— in order to give a position line. Both of these can be given without an allowance for VARIATION so that they can be drawn directly on a chart, and are then respectively called a **QUJ** —

> **"A TRUE heading to steer TO the facility"**

and a **QTE** —

> **"A TRUE heading FROM the facility."** Most VDF-equipped aerodromes offer QDM and QDR only.

Each service is classified according to its accuracy:

Class Alpha is accurate to +/−2°

Class Bravo is accurate to +/−5°

Both the figure and its classification must be read back by the pilot to guard against mishearing. Should the conditions prove unsuitable for accuracy, no information will be given but the operator will give the reason for not doing so.

Aeronautical stations offering VDF are listed in the AIP (COM section). Some stipulate that the service is not available for en-route navigation except in an emergency.

The lesson begins with your aircraft in contact with Scottish Information on 124·9 who must be advised of your intention to contact the VDF facility at Newcastle, as in (1).

In (2) the call is made using the codeword "Homer" which will alert the ATC personnel to your request for VDF information. Note that no initial contact call is necessary. To give the display sufficient time to react to your transmission, you must transmit for more than three seconds.

In (3), the information is read back and the intention to recontact Scottish Information is announced, all in one call, thus saving transmission time.

DIRECTION FINDING — VDF	
YOU — already in contact with Scottish Information — 124·9	**AIR TRAFFIC CONTROL**
(1) Scottish Information, G-JK is changing to NEWCASTLE HOMER 118·5 for DIRECTION FINDING, over.	G-JK is cleared to change frequency to Newcastle Homer 118·5. Advise when back on this frequency.
Will call on rejoining, G-JK.	

(2) Newcastle Homer, this is G-HIJK transmitting for QDM, G-HIJK, over.	G-HIJK, Newcastle Homer. QDM is 170, CLASS BRAVO, over.
(3) QDM 170, class Bravo. Changing back to Scottish Information 124·9, G-HIJK.	G-JK is cleared to change frequency to Scottish Information 124·9.

(4) Scottish Information, this is G-HIJK back on frequency, over.	G-JK, Scottish Information, roger.

Similarly, if you wanted a position line:

(5) G-HIJK transmitting for QDR, G-HIJK.	

Instructor

Whilst this is a relatively simple exercise, it is important that the background information listed above is well understood. When repeating the lesson, cover each side in turn as before until the student can manage without the text. Alter the information given by you in each case, e.g.

"G-JK, Newcastle Homer. Your QDM is 035 Class Bravo." At some point, ask the student to transmit again, to simulate too short a transmission having been made.

Further Study

Read pp101 and 102 CAP413,
UKAIP Vol 2 COM 0-4.

Lesson 9: ATIS: Automatic Terminal Information Service

As traffic increased at certain airfields, the task of providing airfield data to individual arriving and departing aircraft on the tower frequency became impossible. Consequently, some of the larger airfields installed ATIS.

ATIS is a continuously-broadcast tape-recording which provides regularly-updated information on the runway in use, wind, visibility, altimeter settings and temperatures — in short, much of the data that a pilot needs before take-off or landing (1). It can be made available on a dedicated frequency or even via a nearby VOR and so removes the need for tower staff to continually repeat the same information to each pilot.

Every update is designated by a letter in the phonetic alphabet: Alpha, Bravo, Charlie and so on in sequence as conditions alter. After listening to ATIS, pilots then inform the Approach or Ground controller that they have the most recent update in their initial call (2) and (3).

Incoming and outgoing pilots who do not do this may be instructed to refer to ATIS first before calling back and so may be delayed.

Instructor
If possible, ensure that your student has the opportunity of listening to an ATIS whilst on a cross-country flight. Demonstrate the listing of ATIS frequencies in the UKAIP (COM) section.

AUTOMATIC TERMINAL INFORMATION SERVICE	
YOU	**AIR TRAFFIC CONTROL**
(1)	"Manchester INFORMATION ROMEO: Runway in use — 24. Wind 260/17kts, visibility 4000m, RVR on request. QNH998, QFE990, temperature +12, dew point +10. Manchester Information Romeo out.

Example of an inbound aircraft.

(2) Manchester Approach. This is G-IJKL on 119·4.	G-IJKL, Manchester Approach. Pass your message.
G-IJKL is a Piper Malibu, overhead Leek heading 320 2000' VMC estimating CON 25, with Information Romeo.	

Example of a departing aircraft.

(3) Manchester Ground, G-IJKL with INFORMATION HOTEL, request taxi clearance for SVFR flight to Schiphol.	

Lesson 10: Voice Weather Broadcasts (VOLMET)

In addition to the weather information available on request from ATSUs, there is a regularly-updated tape-recorded service also supplying ACTUAL reports, FORECASTS and warnings for the major airports. This service is called VOLMET.

VOLMET is continually broadcast twenty-four hours a day on three frequencies depending upon the location concerned. The airfields are covered in rotation as listed in the UKAIP: (MET 3-1), and take the form of half-hourly weather reports (METARS), at 20 minutes and 50 minutes past the hour.

LONDON VOLMET:

Main — 135·375mHz for Heathrow, Gatwick, Birmingham, Manchester International, Prestwick, Dublin, Amsterdam, Brussels, Charles de Gaulle.

South — 128·6mHz for Heathrow, Gatwick, Luton, Stansted, East Midlands, Southend, Cardiff, Jersey.

North — 126·6mHz for Heathrow, Manchester International, Glasgow, Edinburgh, Prestwick, Belfast, Aberdeen, Newcastle, Leeds/Bradford.

As you can see, the service covers certain airfields outside the UK. Each report is broadcast in the following sequence:

Surface wind
Visibility
RVR if applicable
Weather
Cloud
Temperature
Dewpoint
QNH

In addition, the terms "SNOCLO" and "CAVOK" may be used when appropriate.

Instructor
If possible, when airborne with your student, you should tune in to VOLMET and copy the weather report for a selected airfield.

Further Study
Refer to UKAIP Vol 2, MET 3-1.

VOLMET EXAMPLE	
	frequency 128·6
	East Midlands at 1240 Greenwich: Wind 250/14kts. Visibility 10km or more. 3 octas at 3200′, 6 octas at 6000′. Temperature +13 Dew Point +10 QNH1024.

Lesson 11: The Lower Airspace Radar Service (LARS)

The Lower Airspace Radar Service is an immensely practical service available to all pilots flying up to and including Flight Level 95 outside regulated airspace, and within approximately 30nm of each participating ATS unit.

The service is discretionary and operates either 24 hours a day (H24) or approximate office hours on weekdays. At other times, if no reply is received after three consecutive calls then it can be assumed that the service is not available.

You may be offered one of two services. The first is a RADAR ADVISORY SERVICE. This means that the radar controller will advise you, by bearing and distance, of the whereabouts of traffic appearing on radar to conflict with your intended path. Advice to achieve separation will follow if the traffic is not subsequently sighted. A RAS is obviously of most use to IFR traffic or in conditions of poor visibility.

The second type of service is a RADAR INFORMATION SERVICE. Under this service, you may be given the range and bearing of traffic, but the onus is upon you to see and avoid.

The LARS can only operate within the vertical and horizontal limits of the equipment it uses. If your aircraft is at an inadequate height to give a satisfactory return, or if the controller tells you that you are falling off the edge of his screen and are therefore flying out of range, then he will tell you why he cannot offer you a full service.

Should the aircraft be in IMC, then the pilot may be given an altitude below which full service will not be given. When using the LARS, the QUADRANTAL RULE applies.

Pilots must maintain a listening watch on the frequency allocated and follow the controller's advice or advise him of non-compliance. No change of heading or altitude must be made without prior warning to the controller, who should also be informed when the service is no longer required.

Whenever possible, the pilot will be "handed over" from unit to unit. If so, the new controller may well advise you that he has your details, thus removing the need for you to make a new PHACER call. If you are not handed over, you will be informed of the next ATSU to contact.

Instructor

After the initial run-through, emphasise the pilot's responsibility to obtain clearance when flying through controlled airspace as in Lesson 3.

Further Study

UKAIP Vol 1 RAC 3-6-13.

THE LOWER AIRSPACE RADAR SERVICE	
The aircraft flown is a TB10 Tobago. The frequency changed to is that of Wattisham Radar — 135·2mHz.	
YOU	**AIR TRAFFIC CONTROL**
(1) Wattisham Radar, G-JKLM requesting Lower Airspace Radar Service.	G-JKLM, Wattisham. Pass your details.
(2) G-JKLM is a TB10 VFR Ipswich to Old Warden (P) 4m west of Ipswich (H) heading 160 at (A) 1500′ on the Chatham (C) VMC (R) Request Radar Advisory Service	Roger, G-LM. Are you transponder equipped?
Affirm, G-LM.	Roger, G-LM. Squawk 1771.
"Squawk 1771," G-LM.	

	G-LM, you are identified. Radar Information Service only due to poor radar performance. Maintain VFR and 1500′ on Wattisham QNH of 992. Call overhead Sudbury.
"Maintain VFR and 1500′ on QNH of 992 and call overhead Sudbury," G-LM.	G-LM, that is correct.

(3)	G-LM, you are approaching the Stansted Special Rules Zone. Maintain VFR, — frequency 125·5 and squawk 4321.
Roger. Maintain VFR. Contact Stansted Radar on 125·5 and squawk 4321, G-LM.	

Section C

Lessons 12 – 16
EMERGENCIES, AIRWAYS, CROSSING, THE FRTO TESTS

Emergencies general
Priority scale

Lesson 12: Distress
NIPAHQ call, R/T Silence
monitoring, unacknowledged calls
D&D, cancelling distress
intercepted calls
calls on another's behalf
emergency transponder codes

Lesson 13: Urgency
Limitations, TYRO

Lesson 14: Practice PAN
Requirements

Lesson 15: Radio failure
Types of failure
fault finding
emergency procedures

Lesson 16: Airways crossing
SVFR requirements, entry points

Taking the Tests
The written test, example questions
The simulated flight test, an example of the sort of dialogue you might expect
Regional Directory of Examiners for the FRTO Licence.

Lesson 12: Emergencies

As far as R/T is concerned, there are two categories of emergency, not including Radio Failure which will be looked at later. These are:

1. DISTRESS
2. URGENCY

Top of the PRIORITY SCALE is:

DISTRESS

A situation demanding a Distress (or "MAYDAY") call to be made is one in which there is a "grave and imminent" threat to life, such as an engine failure, a cabin fire or a control failure.

As you can see from the scale below, the Distress call has priority over all other types of transmission.

1. Distress messages and Distress traffic
2. Urgency messages (PAN calls)
3. Direction Finding messages
4. Flight Safety messages (FLIGHT PLANS, company messages of immediate concern, MET advice of immediate concern, e.g. reported WINDSHEAR, anything else of immediate concern).
5. MET messages (Forecasts, observations)
6. Flight Regularity Messages (Loading, timetabling, servicing, amendments, diversions, urgent spare parts, preflight arrangements, company messages relating to movements).

If the call is made on a Tower frequency, ATC may request the aircraft in distress to change to an alternative frequency to prevent interference from other traffic, but once contact is established with a station, no other change of frequency should be made unless acknowledged.

The DISTRESS or MAYDAY call (a)
The call should be made to the station with whom the pilot was last in contact. It begins with the codeword "MAYDAY" spoken three times (1). It is made "all-in-one", i.e. with no Initial Contact Call nor any pause, in order to minimise delay.

The recommended mnemonic to use is NIPAHQ:

N — NATURE of the emergency
I — INTENTION of the pilot-in-command, i.e. "intend ditching
P — POSITION
A — ALTITUDE
H — HEADING
Q — QUALIFICATION

of the pilot-in-command, i.e. state whether

1. Student pilot
2. No instrument qualification
3. IMC rated
4. Full Instrument Rating

This will enable the controller to tailor his advice to your abilities.

The station receiving the call will acknowledge (2) and then clear the frequency for the MAYDAY traffic to use (3). Upon hearing a MAYDAY call, all other stations must immediately cease transmitting and maintain radio silence unless required to render assistance. Should you interrupt a simulated MAYDAY call during the R/T practical test you will probably fail. If you should do so in a real-life emergency you may be jeopardising other people's lives.

Remember that ALL other stations must also monitor the Distress call until it is certain that help is being given. Do not assume that because the call is not addressed to you that it does not concern you. COPY DOWN THE DETAILS.

Unacknowledged distress calls (b)
If your Distress call is not acknowledged then you should announce your intention to change to the International Emergency Frequency of 121·5mHz (4). Note that all calls you now make are prefixed by the codeword "MAYDAY" until the state of distress is cancelled.

Having changed frequency, you should repeat your call, addressing it to the relevant DISTRESS AND DIVERSION CELL ("Drayton Centre" within the London FIR or "Scottish Centre" within the Scottish FIR). Having received your call, "D and D" will acknowledge it and then organise aid (5).

Cancelling a distress call (c)
Should the cause of your distress be removed, then you must cancel your Distress call saying why, and stating your intentions (in real life you would divert to your nearest alternative (6). If you have broadcast an emergency on two frequencies, as in this exercise, then you must cancel your distress call on both.

DISTRESS	
YOU — frequency 124·6	**STATION**
(1) MAYDAY, MAYDAY, MAYDAY. London Information, this is G-KLMN a Pa32 Saratoga (N) with engine failure (I) intend forced landing (P) 10 miles north of Brecon (A) 4000′ (H) heading 030. (Q) IMC rated	(2) G-KLMN, London Information. Roger MAYDAY. BREAK.
(3)	All stations. Stop transmitting. MAYDAY. London Information, out.

If your call is not acknowledged:

(4) MAYDAY. G-KLMN changing to 121·5, out.	
MAYDAY, MAYDAY, MAYDAY. DRAYTON CENTRE, This is G-KLMN A Pa32 Saratoga with engine failure Intend forced landing 10 miles north of Brecon 3000′ Heading 030.	(5) G-KLMN, Drayton Centre. Roger your MAYDAY, out.
(5) MAYDAY, Drayton Centre. This is G-KLMN. CANCEL DISTRESS Engine restarted Diverting to Cardiff.	MAYDAY G-KLMN, Drayton Centre. Understand Cancel Distress, out.
Drayton Centre, G-KLMN is returning to London Information on 124·6 to cancel Distress.	G-KLMN, roger. Frequency change approved.
	MAYDAY ALL STATIONS. This is Drayton Centre. 1313 hours, G-KLMN Distress traffic ended, out.
(6) MAYDAY London Information This is G-KLMN Cancel Distress Engine restarted Diverting to Cardiff.	MAYDAY G-KLMN, London Information. Understand Cancel Distress, out.
	MAYDAY ALL STATIONS. This is London Information. 1315 hours G-KLMN Distress Traffic ended, out.

Intercepted distress calls (d)

If you intercept a distress call which is not acknowledged, such as might occur in the circumstances depicted in the diagram below:

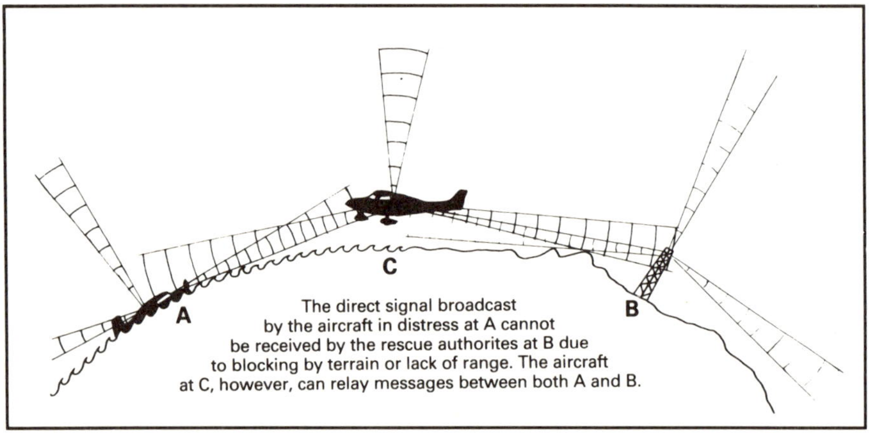

The direct signal broadcast by the aircraft in distress at A cannot be received by the rescue authorites at B due to blocking by terrain or lack of range. The aircraft at C, however, can relay messages between both A and B.

then you must assume responsibility and acknowledge it yourself by relaying the Distress message and any subsequent radio traffic to the station with whom you are in contact. To make it plain just who is in distress, that aspect of the call is repeated (7).

Similarly, you should transmit a MAYDAY call on another's behalf if you should spot someone else in distress, i.e. another aircraft ditching or a vessel on fire.

Secondary surveillance radar and distress

An aircraft in distress not already squawking a transponder code should immediately change to the EMERGENCY TRANSPONDER CODE of 7700, Mode A. An aircraft in distress already squawking a designated code should not automatically change to the Emergency Code but should continue with the transponder code that it has been allocated.

Instructor

It is imperative that this exercise is very thoroughly understood by the student. Success in real life may, as in the practical test, depend upon it.

Further Study

Read thoroughly pp.91-96 in CAP413, COM 0-4.2.3 and the whole of SAR section in volume 2 of the UKAIP. See also AIC 7/1988 (p101) "Use of the VHF Aeronautical Emergency Service by Pilots in a State of Distress, Urgency or in Difficulty".

(7) MAYDAY, MAYDAY, MAYDAY. London Information, this is G-LMNO Have INTERCEPTED MAYDAY from G-KLMN a Pa32, I SAY AGAIN, from G-KLMN with Engine failure making a Forced landing 10 miles north of Brecon 4000′ Heading 030 magnetic. Has changed frequency to 121·5, over.	G-LMNO, London Information. Roger intercepted MAYDAY and frequency change, out.

Lesson 13: Urgency (PAN)

There are four important differences between URGENCY or "PAN" calls, and Distress calls.

(i) An Urgency call is required when a situation which demands urgent attention arises, but where there is NOT YET a threat of "grave and imminent danger". Consequently:
(ii) Urgency calls are second in priority to distress calls.
(iii) There is no specified transponder code for Urgency messages. Finally,
(iv) unlike Distress calls, pilots are encouraged to practice Urgency calls on 121·5mHz. (PRACTICE PAN CALLS), to familiarise themselves with the procedures and the service (see p.48).

Circumstances requiring an Urgency call, as opposed to a Distress call, might be:

a passenger becoming seriously ill,
the pilot having observed a boat adrift with a broken mast,
an aircraft observed to have cartwheeled at a private strip — that is, anything which might lead to a serious threat to life if ignored.

Urgency and Distress calls have the following in common:

(i) they share the same mnemonic, NIPAHQ.
(ii) They share the procedures for cancellation of a call on utilised frequencies and also the procedures for
(iii) the "passing on" of unacknowledged calls.

As with Distress calls, the novice pilot can advise the controller of his inexperience by prefacing his call with the codeword "TYRO".

Aircraft transporting sick or injured passengers or urgently-needed medical supplies can be identified by a PAN MEDICAL CALL. By prefacing his transmission with "PAN" spoken three times and the internationally-recognised codeword "MEDICAL", the pilot can alert the controller to the need for prompt handling.

Instruction
During a flying lesson, perhaps on the way back from a cross-country flight, you should at some time make a Practice PAN call with your student.

Further Study
Read carefully pp.96-98 CAP413.

URGENCY	
YOU – already in contact with London Information on 124·6.	**STATION**
PAN, PAN, PAN, PAN, PAN, PAN London Information, this is G-MNOP, a Mooney 201 (N) I have a passenger on board who has suffered a heart attack. (I) I would like headings to steer to the nearest field at which he can receive medical attention. (P) We are currently coasting in, four miles northeast of Maryport. (A) at 4000′ (H) heading 120 (Q) PPL with no instrument qualification.	G-MNOP, London Information. Roger your PAN call. Can you maintain visual conditions?
Affirm, G-MNOP.	Roger, G-OP. Standby.
	G-OP, the nearest suitable airfield to you is Carlisle, twenty-five, that's TOO FIFE miles to the north-east of your present position. I am in contact with Carlisle to arrange medical help. Contact Carlisle Homer on 123·6 for QDM. They have your details.
Roger, changing to Carlisle Homer on 123·6, G-OP.	G-OP, that is correct.

Carlisle Homer, this is G-MNOP on 123·6. Request QDM. G-MNOP.	G-OP, Carlisle Homer. Your QDM is 072, class Bravo.
QDM 072, class Bravo, G-OP.	G-OP, that is correct. We have your details. Medical help will be here to meet you. Maintain this frequency and call when you have the city of Carlisle in sight.
Roger, call when city of Carlisle in sight, G-OP.	G-OP, that is correct.

Lesson 14: Practice Pan

The London region Distress and Diversion Cell (D&D), is situated at the London Air Traffic Control Centre (LATCC) near Heathrow and is open 24 hours a day, 365 days a year. In addition to providing help and advice to aircraft in distress or genuine urgency, D&D also provide a means by which pilots can familiarise themselves with the procedure for calling for aid — the PRACTICE PAN.

Every pilot should make a Practice PAN call at some time — preferably as a student. This is easily done on a cross-country flight or returning from a lesson.

First of all, decide the nature of your emergency. For example, it might be one in which you require radar vectors in a hurry, as in Lesson 13, or one in which you may be lost as in the example opposite.

Secondly, give yourself some altitude so that you can be identified on radar.

Before you call on 121·5, listen out in case there is a genuine emergency in operation. D&D could be busy and may ask you to try on another occasion.

Be sure to prefix your PAN call with the word "Practice" each time to make it obvious that this is not a genuine urgency call (1).

Your message will be received by one of 6 remote aerials which will indicate the approximate region you're transmitting from. The team of 3 at D&D will immediately respond to the call, contacting DF stations if necessary and plotting your position manually on a large wall map in front of them. If necessary, large-scale O.S. maps will be used to identify ground detail to help the pilot, as in (3).

D&D's radar images are computer-generated and selectively respond to the emergency transponder codes. If one of these codes is used, then nearby traffic controllers will also notify D&D. To avoid causing unnecessary concern do not use the emergency transponder codes unless agreed.

For example, if you arrange beforehand, you could practise using the emergency code 7600 (radio failure). D&D's response in this case could be:

Pilot	D&D
	"Aircraft in (area) squawking radio failure, change squawk to (7 - - - -) if you are receiving me."
(Pilot does so).	
	"Squawk (7 - - -) if you have more than half an hour's fuel remaining."

And so on.

In this way, the nature of the Practice Pan (or indeed real urgency), can be identified and help given.

PRACTICE PAN	
YOU – on frequency 121·5	**DRAYTON CENTRE**
PRACTICE PAN, PRACTICE PAN, PRACTICE PAN. Drayton Centre, this is G-NOPQ requesting PRACTICE PAN.	G-NOPQ, Drayton Centre. Understand you want a PRACTICE PAN. Standby. I'll call you back in two minutes.

(1)	G-NOPQ, Drayton Centre. Go ahead with your PRACTICE PAN
PRACTICE PAN, PRACTICE PAN, PRACTICE PAN. Drayton Centre. This is G-NOPQ, a Cessna P210 VFR Carlisle to Newcastle. We are lost We'd like headings to steer for Newcastle Last known position was 10 miles east of Kielder Dam 2000′ heading 110 no instrument qualification	G-NOPQ, this is Drayton Centre Roger your Practice Pan call. Can you maintain visual conditions?
Affirm, G-NOPQ	Roger, G-PQ. Maintain present heading and keep clear of cloud. Standby to transmit for Direction Finding. When instructed, I want you to transmit your callsign and count to five slowly. Do you understand?
Affirm, G-PQ. (2)	Roger, G-PQ. Transmit for Direction Finding now.
G-NOPQ, 1, 2, 3, 4, 5.	Thank you, G-PQ. Standby
(3)	G-PQ, your position is 7 miles north-west of Newcastle airport. Do you see a disused airport in your two o'clock?
Affirm, G-PQ	G-PQ, that is the disused aerodrome at Morpeth. Turn now onto a heading of 140 for Newcastle, range seven and a half miles, and call Newcastle Approach on 126·3.
(4) That coincides exactly with our position. G-PQ is cancelling the Practice Pan and changing to Newcastle Approach 126·3.	

D&D encourage the Practice Pan service. They also encourage organised visits

O/C LATCC (Mil),
T & A Squadron,
RAF West Drayton,
Porters Way,
West Drayton,
Middlesex.
Tel: 0895-444077 ext 3395.

If you do organise a visit to LATCC you might like to also organise a pilot friend to make a Practice Pan call to coincide with it.

One last note, don't make a PRACTICE Pan call if you are in genuine need of help just to save an imaginary embarrassment. If a real PAN is in progress you will be refused and then you will have to own up. If you need help then don't delay until you are in a fix with little fuel or daylight remaining. Make a proper Pan call early. There are no comebacks. Nobody at D&D will make a judgement. Don't wait.

Further Study
UKAIP Vol 2, SAR.

The above text also applies to the Scottish FIR.

Lesson 15: Radio Failure

Total and irreversible radio failure is a rarity. More often than not, the "radio failures" we suffer are caused by faults in the peripheral equipment, or are temporary gremlins in either transmitter or receiver. It's very seldom that both fail simultaneously, and often it's possible to find a way around the "radio failure" if we can determine its nature and extent.

All "radio failures" fall into the two categories of "apparent" radio failure and "actual" radio failure, i.e.

(a) those which can be resolved without affecting the flight and
(b) those which appear unsolvable during flight and which demand that a specific course of action be taken.

Heading the list of "apparent" radio failures are those arising from radio management, popularly known as "finger trouble". Everyone at some stage makes the mistake of setting a similar-but-incorrect frequency or of not fully inserting a jackplug, or of turning the speaker volume down for a moment and then forgetting to turn it up again. And a common error where radio installations have more than one radio "box" is to have the audio panel set up wrongly.

If your pre-flight Radio Check hasn't met with any response then it could be that your transmission is screened from the tower by hangars or other large buildings, or that you have made the check before starting the engine and the battery power is insufficient. Physical blocking of messages can also occur in flight, usually by high terrain. Moreover, the "line-of-sight" nature of VHF transmissions may mean that the earth's curvature prevents a station receiving your messages, especially if your aircraft is at a relatively low level. Other aircraft may be able to assist in this event by relaying messages (see Distress).

It could simply be that the station you are calling is closed or unmanned, in which case London Information or Scottish Information as appropriate may be able to help by contacting that station by landline on your behalf if so requested.

The majority of technical problems causing "apparent" radio failure can be resolved by using a "diagnostic" approach. Once sure that "finger trouble" isn't the cause, the pilot should ask himself some questions — no, not:

> "What on earth am I doing here when I could be safe at home in bed?" but:

1. IS THE RADIO INSTALLATION AS A WHOLE LIFELESS?

If is it is a combined NAV/COM installation, check that the NAV side is working. If your radio has an LED display, check that the frequency is plainly visible. Check that you can hear SIDETONE when you transmit (i.e. Can you hear yourself speaking?) and that you can receive continuous broadcasts such as VOLMET or ATIS.

Check that the BATTERY/MASTER switch is ON.
IF NOT — Set to ON.

Check that alternator and radio CIRCUIT BREAKERS are set.
IF NOT — Reset. Thermal trip breakers are installed in the circuit to guard against overload or fire damage. Once allowed to cool for two minutes they may be reset. If a circuit breaker trips again it could indicate a fault in the system requiring ground attention rather than a "nuisance" trip-out.

Check that the set's ON/OFF switch is functioning.
RECYCLE ON/OFF switch and SET.

Check ammeter doesn't show a discharge.
Switch radios OFF and shed any unnecessary load, RECYCLE both sides of the battery/master switch and then switch back to ON. If a DISCHARGE or a LOW voltage still shows (CHARGING FAILURE), then minimise any current drain on the battery, as it will only supply power for a limited period. You may wish to reserve power for flaps, landing lights or a radio call when closer to your alternate aerodrome. Consult PILOT'S MANUAL for specific action to take in your aircraft. If the radio still appears to be lifeless then assume that you have an actual receiver failure. Squawk CODE 7600 and TRANSMIT BLIND. (This has nothing to do with poking your eye out, see following example). If the receiver IS working, then ask:

2. IS THE TRANSMITTER AT FAULT?
After announcing your intention to change, try a radio check on another frequency.

Check
With headphones on, can you hear sidetone?
IF NOT — Change to HANDMIKE or use EPTT button if fitted.

If you suspect transmitter failure, then listen out on the designated frequency anyway as the ground station may be able to receive your CARRIER WAVE and be able to communicate with you as in the following example.

The aircraft being flown is G-QRST and is already in contact with Swansea Approach on 119·7mHz.

YOU	AIR TRAFFIC CONTROL
(Your return-to-field call is made but there is no immediate response).	"Station calling Swansea Approach on 119·7, I am receiving you on carrier wave only. If that is G-QRST, confirm by answering "yes" with a one-second press on your transmit button. If not, use two presses to indicate "no". Is that G-QRST?
(Press transmit button for one second).	Roger, G-ST. Are you returning to the field from the eastern training area?
(Press transmit button for one second).	G-ST, that's understood. Transmit four times when you have the field in sight. Expect to join downwind for a left-hand approach to runway 22, QFE999. Do you understand?
(Press transmitt button for one second).	

The return to the field could be made like this with the controller issuing traffic-avoidance, landing and taxi instructions until the flight is terminated. Although other traffic would be advised of your non-radio join, it would still be necessary to keep a good look-out, and to watch for light signals from the tower. Such an arrangement works well at smaller airfields in VMC. However, radio failure when flight planned or cleared into a busy field in controlled airspace is a complicated matter. Much depends upon weather conditions, type of failure, type of clearance and ETA, and many fields have standing instructions peculiar to that field. If you plan to fly into an airfield in controlled airspace it is your responsibility to discover that airfield's particular procedure.

Further Study

AIC 111 (p127) 89 "Radio and Navigational Failure in Flight".
UKAIP Vol 1: RAC 6.3 to 6.6.

Should you have the double misfortune of losing voice transmission (i.e. carrier wave only), during an emergency (for example if you became lost), then you can use a similar procedure on the emergency frequency of 121·5mHz. First, you must attract the attention of the emergency services by pressing the transmitter button four times to transmit the Morse letter "H". Upon receiving this, the controller will understand that you require a homing. Transmit "H" at ten-second intervals to allow time for VDF (see Urgency). If circumstances deteriorate and you face an increasing degree of difficulty then you can indicate this with the Morse letter "X" (–··–).

3. IS YOUR RECEIVER AT FAULT?

CHECK that you can hear background noise, using the squelch control or the test switch.

CHECK that you can hear continuous broadcasts such as VOLMET or ATIS.

IF NOT — change headsets or use a mike/speaker combination.

Suspect receiver failure.

If you suspect receiver failure, assume nonetheless that the transmitter is working and transmit "blind" at the scheduled times and positions as in the following example. Note that the message is given twice.

In this example, failure is suspected after the initial contact call to inform the Approach controller that the aircraft is inbound.

YOU	
"Liverpool Approach, G-QRST, transmitting blind on 119·85 due to receiver failure. I have the field in sight and will advise when joining right-hand downwind to 09. We have the Cessna on the crosswind leg in sight. I say again, Liverpool Approach, G-QRST, transmitting blind on 119·85 due to receiver failure. I have the field in sight and will advise when joining right-hand downwind to 09. We have the Cessna on the crosswind leg in sight."	

At this point, the controller would inform and advise his traffic accordingly.

"Liverpool Approach, G-QRST, transmitting blind due to receiver failure. Am joining right-hand downwind for runway 09, in visual contact with the Cessna ahead." I say again, "Liverpool Approach, G-QRST, transmitting blind due to receiver failure. Am joining right-hand downwind for runway 09, in visual contact with the Cessna ahead."	

Similarly, G-QRST would transmit blind at the normal reporting points remaining.

In such a situation, the returning pilot must keep a sharp lookout for light signals from the tower and for other traffic.

EMERGENCY TRIANGLES

The pilot with a suspected transmitter failure whilst in need of assistance — for example if lost above a total cloud cover, can also attract the attention of a ground radar authority by flying an equilateral triangle pattern of two minutes for each leg, flown at Best Endurance Speed, with turns made as tightly as possible and allowance made for Wind Effect upon track and length of leg flown. Upon recognising your pattern, and remember that you must have adequate height for radar tracking, the controller will attempt to contact you on the emergency frequency. You should listen out on 121·5 even if you believe your receiver to be useless.

If the controller sees no response to his heading instructions then he will assume that you have receiver failure and he will attempt to put up a shepherd aircraft to lead you into clear conditions and to a suitable airfield. Aircraft in your vicinity may also be asked to contact you so you must switch on all lights and keep a good lookout. Flying the triangle **CLOCKWISE** automatically indicates that your transmitter is believed to have failed but your receiver is operating:

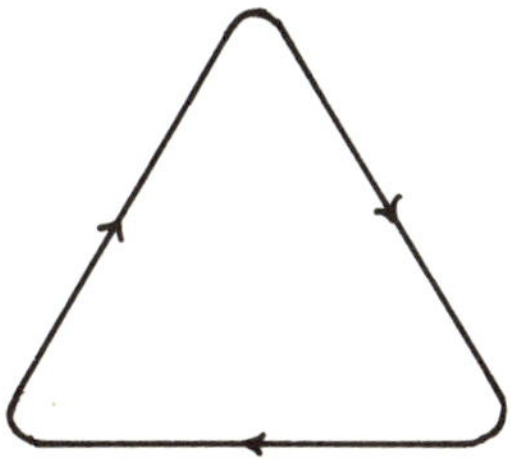

Clockwise
If only your receiver is operating

Flying **ANTICLOCKWISE** indicates the suspected failure of both transmitter and receiver. Note the use of the word "suspected" for, with your receiver silent, you would not know if your transmitter was working.

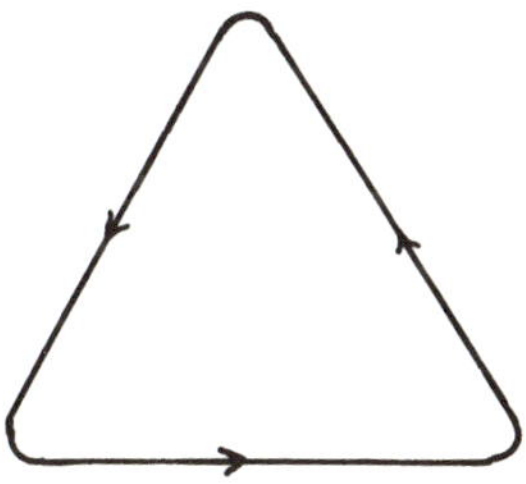

Anticlockwise
If NEITHER transmitter nor receiver is operating

In both cases, light aircraft should fly each leg for two minutes.

Remember

a. Maintain a safe margin for terrain clearance.
b. Remain clear of controlled airspace.
c. Fly an accurate, recognisable triangle. On the radar screen, it could otherwise appear as a circle or orbit.
d. Fly at least two triangles before resuming heading. Then repeat.
e. Remain in VMC if possible.
f. This procedure is a "last resort" and does not replace the transponder codes 7600 and 7700.

Further Study

AIC 111/1989 (p127) 7 Dec.

Lesson 16: Airways Crossing

If he's already filed a flight plan and if he makes the initial call ten minutes before ETA at his crossing point, then the pilot of a light aircraft will be permitted to cross an airway at levels higher than the Flight Level "floor" of that airway.

The aircraft in this example has an American registration. As with UK registrations its abbreviated form is only used after the callsign has been shortened by ATC.

Following the initial contact call (1), comes a modified version of our old friend the PHACER call (2). After obtaining his clearance, the pilot will be asked to make a position report when entering airways, (3) and also when leaving, (4).

Instructor

Once your student is familiar with the format of the calls, ask him to compose further Airways Crossing requests using a topographical map or an AERAD chart.

Further Study

UKAIP Vol 1 RAC 3-5.

AIRWAYS CROSSING	
YOU – on frequency 133·05	**AIR TRAFFIC CONTROL**
(1) (R) Manchester CONTROL, this is N63N08J for CROSSING CLEARANCE of B1 at Lynas, over.	N63N08J, Manchester Control. Pass your details.
(2) N63N08J is a Cessna 404 (P) overhead Harlech (H) heading 357 (A) FL65 (C) VMC (E) to cross at Lynas, FL65, time 40.	Manchester Control clears N63N08J to cross B1 at Lynas. Maintain FL65 whilst in controlled airspace.
"Cleared to cross B1 at Lynas. Maintain FL65," N63N08J	N08J, that is correct. Report entering airways.
"Report entering airways," N08J.	

(3) Manchester Control, N08J is entering B1 at Benllech, FL65.	Roger, N08J. Report leaving airways.
Will report leaving airways, N08J.	

(4) N08J is leaving airways 5nm north of Lynas, and changing to Ronaldsway Approach on 120·8.	N08J, roger. Frequency change approved.

The RTF Written Test

The RTF written test comprises:
eighteen multiple-choice questions
one "written" answer.

It may be conducted by a PPL(GR), a PPL(X), or an approved RTF examiner.

The following is a sample of the sort of questions asked in the multiple-choice section.

1. What does "roger" mean?
(a) Yes
(b) I have received all of your last transmission
(c) That is correct

2. What is the abbreviation for the callsign Avia G-WXYZ?
(a) Avia G-YZ
(b) Avia YZ
(c) G-YZ.

3. What is readability 4?
(a) Readable
(b) Readable now and then
(c) Perfectly readable

4. How accurate is a Class B VDF bearing?
(a) +or− two degrees
(b) +or− five degrees
(c) +or− one degree

5. How would you transmit 1500 feet?
(a) One Thousand Five Hundred
(b) Fifteen Hundred Feet
(c) One Five Zero Zero

The nineteenth question requires a written answer.
For example:

19. Assuming that you have sufficient time to include the required information, write down IN THE CORRECT ORDER an example of a distress call and message.

The Simulated Flight Test or "Practical"

The practical communications test must be carried out by an authorised RTF examiner and requires the use of an RTF simulator. During the test, you will be required to make the RTF calls that would be necessary during a simulated cross-country flight, which would include emergency procedures.

There is no longer any requirement to demonstrate airways procedures (see Section D on Airways).

You will be briefed to expect an emergency during the "flight" and will be told how to be notified of this emergency, usually by a warning light on your simulator unit. You may also hear distress traffic. Upon hearing a distress message, you MUST maintain radio silence unless your assistance is required, e.g. to relay that message or any subsequent information. You must not make any change of frequency unless it is evident that assistance is being provided.

On the following pages, you will find a candidates' brief and chart. You will also find a "transcript" of the sort of dialogue that you might expect during your practical communications test.

Candidates' Brief

ROUTE No. 1
Flight Planning details

Aircraft Type: Cessna 152 Callsign G-WXYZ
Cruising Speed: 95kts TAS

Destination BRISTOL Alternate: STAVERTON

Equipment: Single VHF/RTF, VOR, ADF, Transponder 4096 codes (no Mode C)

Pilot Qualifications: PPL/No instrument qualification

Flight Plan		Airborne 1215			
From	To	Hdg	Dist	Time	Altitude
SHOREHAM	MIDHURST MAST	325	19	14	2500′
MIDHURST MAST	BASINGSTOKE	325	21	15	2500′
BASINGSTOKE	BATH	290	47	27	2500′
BATH	BRISTOL	290	14	8	2000′
ALTERNATE					
BATH	STAVERTON	026	32	25	2500′

Instructions
1. The test starts assuming that you have just entered the aircraft which is in the parking area.
2. Carry out the "flight" as indicated on the chart, transmitting all necessary RTF messages on the appropriate frequencies.
3. When approximately 20 miles northwest of Shoreham obtain D/F assistance.
4. Obtain clearance to pass through Farnborough MATZ.
5. Whilst working London Information, after leaving Basingstoke you experience a loss of engine power, make an appropriate urgency call.
6. Obtain Bristol weather and, if suitable, carry out the procedure for approach and landing. If not, divert to Staverton (Gloucester/Cheltenham).
7. At sometime during the "flight", you will be made aware of an emergency. You will be expected to carry out the appropriate RTF procedure until such time exists that you are made aware that the emergency has ceased to exist.
8. It is emphasised this is a radio test. It is NOT a test in navigation or airmanship and it is NOT conducted in real time.

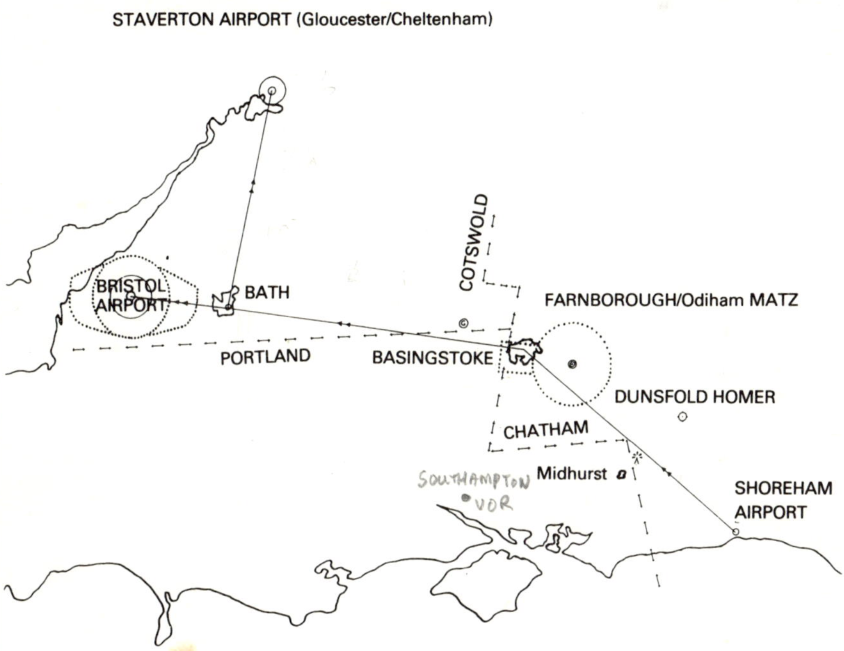

STATION	CALLSIGN	FREQUENCY
SHOREHAM	Tower	125·4
	Approach	123·1
DUNSFOLD	Homer	122·5
FARNBOROUGH	Approach	125·2
BRISTOL	Tower	120·5
	Approach	132·4
STAVERTON	App/Twr	125·6
LONDON	Information	124.7

FRTO ROUTE 1 — SHOREHAM TO BRISTOL	
(1) Shoreham Tower, this is G-WXYZ. Radio check on 125·4, over.	G-WXYZ, Shoreham Tower reading you fives.
(2) Reading you fives also, request airfield data and taxi clearance for VFR flight to Bristol, G-YZ.	G-YZ, roger. Taxi to holding point Bravo for runway 03, QNH1010.
Taxi to Bravo for runway 03, QNH1010, G-YZ.	G-YZ, correct.

(3) G-YZ is ready for departure.	Roger. G-WXYZ is cleared for take-off, left turn out on track, wind 030 degrees, 12 kts.
"Cleared for take-off, left turn out on track," G-YZ.	G-YZ, correct.

	G-YZ airborne 05. Change now to Approach on 123·1.
(4) Changing now to Approach on 123·1, G-YZ.	

Shoreham Approach, this is G-WXYZ on 123·1, over.	G-YZ, roger. What altitude will you be climbing to?
G-YZ will be climbing to 2500′.	Roger, call when you are clear of the zone.
Roger, G-YZ.	

G-YZ is clear of the zone and changing frequency to London Information on 124·7.	G-YZ, roger, London Information on 124·7.

London Information, this is G-WXYZ on 124·7 over.	G-WXYZ, London Information. Pass your message.
(5) G-WXYZ is a Cessna 152 from Shoreham to Bristol via Basingstoke (P) 5 miles north-west of Shoreham (H) heading 325 (A) 2500′ (C) VMC (E) estimating Basingstoke 34 (R) request traffic information.	G-YZ, roger. No known traffic to affect you. Chatham QNH is 1012.
Chatham 1012, G-YZ.	

G-YZ request frequency change to Dunsfold Homer on 119·8 for direction finding.	Roger, G-YZ. Frequency change approved. Report back on this frequency.
Wilco, G-YZ.	

(6) Dunsfold Homer, this is G-WXYZ on 119·8 for QTE. G-WXYZ, over.	G-WXYZ, Dunsfold Homer. Your QTE 230, class Bravo.
QTE 230 class Bravo. Changing to London Information on 124·7, G-WXYZ out.	

London Information, G-WXYZ is back on 124·7.	G-YZ, roger.

G-YZ changing frequency to Farnborough Approach on 125·2 for MATZ penetration.	Roger, G-YZ. Frequency change to Farnborough Approach on 125·2. Portland QNH 1014.
Portland 1014, G-YZ out.	

(7) Farnborough Approach, this is G-WXYZ for MATZ penetration, over.	G-WXYZ, Farnborough Approach. Pass your details.
G-WXYZ is a Cessna 152 Shoreham to Bristol, turning at Basingstoke 5 miles southeast of your zone heading 325 level at 2500′ on the Portland QNH VMC squawking 4321.	G-YZ, squawk 0741.
(8) Squawk 0741, G-YZ.	

	G-YZ, you are identified. Radar Information Service only due to poor radar return.
Roger, Radar Information Service only, G-YZ.	

	G-YZ, is cleared into the zone, to maintain 2500′ on the Farnborough QFE of 1012. Report entering the zone.
"Cleared MATZ penetration Maintain 2500′ on QFE 1012," G-YZ.	G-YZ, correct. Report entering the zone.
"Report on entering the zone," G-YZ.	

G-YZ is entering the zone.	G-YZ, roger. Fast jet traffic in your two o'clock 2000′ moving right to left. Do you have visual contact?
Affirm, G-YZ.	G-YZ, roger. Call leaving the zone.
"Call leaving the zone," G-YZ.	

G-YZ is leaving the zone to the west and turning on route for Bristol.	Roger, G-YZ. Portland 1014.
Portland 1014, G-YZ.	

	G-YZ, you are approaching the limit of my radar coverage. Radar Information Service Terminated. Squawk 4321. Suggest call London Information on 124·7.
"Squawk 4321, London Information on 124·7," G-YZ.	G-YZ, correct.

London Information, this is G-WXYZ on 124·7, over.	G-WXYZ, London Information. Pass your details.
G-WXYZ is a Cessna 152 from Shoreham to Bristol via Basingstoke (P) 4 miles northwest Basingstoke (H) heading 290 (A) 2500′ on 1014 (C) VMC (E) estimating Bristol 26 (R) request traffic information.	G-YZ, roger. No known traffic to affect your flight at 2500′. Portland QNH1014. Report Bath.
Portland 1014. Report Bath, G-YZ.	

(During the "flight", you will be made aware of an emergency, possibly by the illumination of a "warning light" on your RTF simulator. You will have been briefed on the nature of the emergency beforehand. In this case, it is a temporary partial loss of power.)

PAN, PAN, PAN. PAN, PAN, PAN. London Information, this is G-WXYZ, a C152 (N) engine losing power, (I) intend making a precautionary landing on a gliding site south (P) of Hungerford, (A) descending from 2500′, (H) heading 340, (Q) no instrument qualification, over.	G-WXYZ, roger your Pan.

(10) London Information, PAN G-WXYZ Cancel PAN. Engine power fully restored. Climbing to 2500′, over.	G-WXYZ, roger your cancelled PAN. Report reaching 2500′, Cotswold QNH1012
"Report reaching 2500′, Cotswold 1012," G-WXYZ.	G-YZ, correct. Break. PAN all stations London Information time 44 urgency traffic G-WXYZ ended out.

Level at 2500′, G-YZ.	G-YZ, roger. Report Bath in sight.
"Report Bath in sight," G-YZ.	

(11) G-YZ has Bath in sight, request Bristol weather.	G-YZ, Bristol weather at 1345, surface wind 330, less than 5 kts, visibility 1100m in mist, nil weather, 4 octas 300′, 8 octas 700′, QNH1015 QFE993mb, runway 09.
Weather copied, QNH1015, QFE993, runway 09. Standby, G-YZ.	

G-YZ requests Staverton weather.	G-YZ, roger. Standby.

	G-YZ, are you ready to copy Staverton weather?
Affirm, G-YZ.	G-YZ, Staverton weather at 1345: surface wind 020/12kts visibility plus 10km recent rain showers 3 octas 2000′ QNH1015 QFE1012 runway 36
Weather copied, QNH1015, QFE1012, runway 36. G-YZ is diverting to Staverton.	G-YZ, roger. Call Gloucester in sight.
"Call Gloucester in sight," G-YZ.	

G-YZ has Gloucester in sight.	G-YZ, roger. Call now Staverton Approach on 125·6.
Changing to Staverton Approach on 125·6, G-YZ.	

Staverton Approach, G-WXYZ on 125·6.	G-WXYZ, Staverton Approach. Pass your details.
G-WXYZ is a Cessna 152 Shoreham to Bristol diverting to Staverton (P) 5 miles south of Gloucester (H) heading 025 (A) leaving 2500′ for 1500′ (C) to maintain VMC (E) estimating the field at 26 (R) request joining instructions.	G-YZ, roger. Join straight-in for runway 36. QFE1012.
"Join straight-in for 36, QFE1012," G-YZ.	G-YZ, correct. Call field in sight.
"Call field in sight," G-YZ.	

G-YZ has the field in sight.	G-YZ, roger. You are cleared to final number 1. Call finals.
"Call finals," G-YZ.	

(12) G-YZ final approach to 36.	Y-YZ cleared to land, wind 020/12kts.
"Cleared to land," G-YZ.	

Regional Directory of Examiners

ENGLAND

South-West

AVON, Bristol Airport: Mr. J. Craw (VHF & HF)	0934-514378
CORNWALL, Liskeard: Mr R. E. W. Smerdon (VHF)	020882-419
DEVON, Exeter Airport: Ms M. Farmer (VHF)	0392-67653
DORSET, Bournemouth/Hurn: Mr R. J. Smith (VHF & HF)	0202-499888
GLOUCS, Staverton Airport: Mr J. T. Cole (VHF)	0452-713924
GLOUCS, Staverton Airport: Ms J. A. Lyons (VHF & HF)	0452-712388
WILTSHIRE, Compton Abbas: Mr A. J. Curtiss (VHF/HF)	074781-1767
WILTSHIRE, Marlborough: Mr K. Hope	0672-513614

Central Southern

BERKS, Newbury: Mr T. Orchard (VHF & HF)	0635-60210
BERKS, Maidenhead: Mr Neil Popkin (VHF)	0628-823272
BUCKS, Marlow: Mr J. Deane (VHF)	0494-445412
HAMPSHIRE, Winchester: Mr J. E. Davies (VHF)	0703-643707
HAMPSHIRE, Lasham Aerodrome: Mr M. A. Hook (VHF)	025-683-270/332
HANTS, Old Sarum: Mr B. J. Dyke (VHF)	0722-23385
HANTS, Southampton Airport: Mr R. J. Smith (VHF & HF)	0703-613592/3
MIDDX, Leavesden Aerodrome: Mr S. Awad (VHF & HF)	0923-671411
MIDDX, Leavesden Aerodrome: Mr T. Miller (VHF & HF)	0923-671411
OXFORD, OATS: Captain M. D. Baker	0865-841234
OXFORD, OATS: Captain T. W. Morris (VHF & HF)	0865-841234
OXFORDSHIRE, Bicester: Mr M. J. Orr (VHF)	0869-243030
OXFORD, OATS: Captain P. F. C. A. Thorn (VHF)	0865-841234
SURREY, Blackbushe: Mr L. E. Adlington (VHF)	0252-873747/872152
SURREY, Chobham: Mr R. M. Trott (VHF & HF)	09905-8075
SURREY, Chobham: Mr R. Welfare (VHF)	09905-8502

Southeast

BEDS, Luton: Mr M. J. Lyons (VHF)	0582-24426/36919
BEDS, Luton Airport: Mr R. D. Harwood (VHF)	0582-405250
CAMBS, Bourn Aerodrome: Mr R. Boyer	09544-602
CAMBS, Hadstock: Mr E. Ford (VHF & HF)	07875-3943
CAMBS, Peterborough: Mr J. Nicholls (VHF)	08324-289
CAMBS, Sawston: Mr J. L. Stephens (VHF)	0603-46379/737323
CAMBS, Waterbeach: Mr P. A. C. Stephens (VHF)	0223-862769
ESSEX, Andrewsfield: Mr D. D. Delaney (VHF)	037-86500
ESSEX, Colchester: Mr E. Ford (VHF & HF)	07875-3943 ext 884
ESSEX, Southend: Mr D. Woolf (VHF)	0702-545198
ESSEX, Southend: Mr E. A. Clack (VHF)	0702-545420
ESSEX, Stansted Airport: Mr N. J. Harris (VHF & HF)	0279-815579
HERTS, Elstree Aerodrome: Mr D. T. Hoy (VHF & HF)	081-953-4343
HERTS, Elstree Aerodrome: Mr T. Sharland (VHF)	081-953-4343
HERTS, Panshanger Aerodrome: Mrs S. M. Hewish (VHF)	070-335021
KENT, Biggin Hill: Mr K. D. Allen (VHF)	0959-971919
KENT, Biggin Hill: Mr G. A. Dobson (VHF & HF)	0959-75088
KENT, Manston: Mr A. J. Bradshaw (VHF)	0843-89656/89520
KENT, Rochester Airport: Mr J. Stevens (VHF)	0634-816340/861682
SURREY, Redhill: Ms Jill S. Devlin (VHF & HF)	073782-2959
SUSSEX, Chichester Aerodrome: Mr R. Hutchinson (VHF & HF)	0243-531516
SUSSEX, Chichester Aerodrome: Mr I. B. Sandon (VHF)	0243-774656
SUSSEX, Hove: Mr P. K. Ghose (VHF & HF)	0273-462457/417859
SUSSEX, Shoreham Airport: Mr M. E. K. Peal (VHF & HF)	07917-62277

Central

BEDS, Cranfield: Mr M. Dry (VHF & HF)	0234-750197
BEDS, Cranfield: Mr R. G. Walker (VHF & HF)	0234-750592
BEDS, Cranfield Aerodrome: Mr J. A. Hinwood (VHF & HF)	0234-751243
BIRMINGHAM, B'ham Aero Centre: Mr A. J. Noone (VHF)	021-743-8124
BIRMINGHAM, B'ham Av'n: Mr S. L. Williamson (VHF)	021-742-4509
HEREFORD, Shobdon Aerodrome: Mr D. I. Davies (VHF)	056-881-723

NOTTS, Tollerton Aerodrome: Mr D. Robinson (VHF)	0602-811327
SHROPSHIRE, Sleap: Mr L. H. Phillips (VHF & HF)	0939-32882
WARKS, Coventry Airport: Mr P. H. Smerdon (VHF & HF)	0203-305721
WARKS, Shipston-on-Stour: Mr P. Bryan (VHF)	0608-62167
WARKS, Wellesbourne/Mountford: Mr A. T. Hooper (VHF)	0203-455840

Eastern

LINCS, Skegness Aero Club: Mr D. Coulson (VHF)	0754-2240/66621
LINCS, Fenland: Mr H. Lomax (VHF)	040-634-461
LINCS, Wickenby: Mr R. G. Simpson (VHF)	06734-345/6
NORFOLK, Shipdham: Mr M. J. Brown (VHF)	0362-820337
NORFOLK, Swanton Morley Aerodrome: Mr R. Goff (VHF)	036-283-274/581
SUFFOLK, Ipswich Airport: Mr P. Lucas (VHF)	0473-713312
SUFFOLK, Ipswich Airport: Mr W. Stitt (VHF)	0473-714840

Northern

CHESHIRE, Chester: Mr D. A. Williams (VHF & HF)	0244-535333 ext 8
DERBYS, Kegworth: Mr J. Kinsella (VHF & HF)	0302-535666
CO DURHAM, Darlington: Mr J. L. Povey (VHF)	032-533-2195
HUMBERSIDE, H'side Airport: Mr F. Morgan (VHF)	0652-688859
HUMBERSIDE, H'side Airport: Mr G. C. Stewart (VHF)	0652-688833
LANCS, Blackpool: Mr F. W. Jefferson (VHF)	0253-41871/8868864
LANCS, Liverpool Airport: Mr A. Dickinson (VHF & HF)	051-427-3300
LANCS, Liverpool Airport: Mr N. J. Sutton (VHF)	051-427-7449
LANCS, Preston: Mr G. Hobson (VHF)	09952-4694
LANCS, Southport: Mr W. Raistrick (VHF)	051-427-7739
LANCS, Blackpool: Mr K. R. Whyham (VHF & HF)	0253-45396
MANCHESTER, Manchester Airport: Mr K. Edmunds (VHF)	061-489-3139
MANCHESTER, M/cr Airport: Mr P. Eite (VHF)	061-499-1444
MANCHESTER, M/cr Airport: Mr J. T. Nuttall (VHF & HF)	061-489-3139
MANCHESTER, Barton Aerodrome: Mr P. Dean (VHF)	061-789-4785
NOTTS, Netherthorpe Aerodrome: Mr R. G. Cooper (VHF)	0909-475233
YORKS, Leeds/Bradford Airport: Mr M. Dobson (VHF)	0532-501401
YORKS, Leeds/Bradford Airport: Mr J. H. Fenton (VHF)	0532-503840
YORKS, Sherburn-in-Elmet: Mr K. R. Jackson (VHF)	0977-682674

Isle of Man

CASTLETOWN, Derbyhaven: Mr A. Gardner (VHF)	0624-823454

CHANNEL ISLANDS

JERSEY, Jersey Airport: Mr G. A. Tipping (VHF & HF)	0534-43990

SCOTLAND

ABERDEENSHIRE, Inverurie: Mr G. L. Greavey (VHF & HF)	0224-770332
AYRSHIRE, Prestwick Airport: Mr I. Weir and Mr G. Austin (VHF)	0292-671022
AYRSHIRE, Prestwick Airport: Mr E. Broughton (VHF)	0292-671022
AYRSHIRE, Prestwick Fl. Gp: Mr R. Young (VHF & HF)	0293-79800 ext 2763
CUMBERNAULD, Cumbernauld A'drome: Mr D. A. Smellie (VHF)	0236-722100
FIFE, Fife Airport: Mr B. R. Bailey (VHF)	0592-753792
FIFE, Glenrothes Aerodrome: Mr R. D. Malcolm (VHF)	0382-644577/644372
PERTH, Perth Aerodrome: Mr W. D. B. Andrews (VHF)	0738-5231 ext 222
RENFREW, Renfrew Flying Club: Mr T. J. Goodwin (VHF)	041-887-3398
SHETLANDS, Unst Aerodrome:	095781-541

WALES

GLAM, Swansea Airport: Mr J. S. Small (VHF)	0792-297883

NORTHEN IRELAND

BELFAST, Ulster Flying Club: Mr A. D. Patterson (VHF)	0232-658333
BELFAST, Belfast Airport: Mr M. Woodgate (VHF & HF)	08494-52789
DERRY, Eglington Aerodrome: Mr A. F. Danton (VHF)	0504-810962/8959

Directory compiled from information kindly supplied by CAA.

Section D

Lessons 17 & 18
AIRWAYS PROCEDURES

NOTE: Airways radio procedures will not be tested for the FRTO licence. This section is for the guidance of those FRTO holders who wish to fly on airways. Assessment will be made during the initial instrument flight test.

Lesson 17: Clearances and Position Reports

Flight along airways requires additional radiotelephony techniques, specifically for airways clearances, position reports, and for entering and leaving airways.

AIRWAYS CLEARANCES must have all the essential elements read back verbatim. Initially, many people find this difficult until they realise that the information in the clearance is given in a set order. A typical airways clearance is shown opposite in (1), and is in the following sequence:

(NAME of) Control clears
(aircraft CALLSIGN) to
(DESTINATION) via
(AIRWAY designator), joining airways at
(LOCATION — usually a reporting point overhead a beacon) at
(FLIGHT LEVEL), reporting
(REPORTING POINTS), to squawk
(TRANSPONDER CODE) and to contact
(NAME of) Control on
(FREQUENCY) when cleared.

Not all clearances are as straightforward as in the example opposite. Additional frequencies and airways are sometimes included, but the order is much the same. Once you know what to expect, and the order to expect it in, then airways clearances should hold few surprises. Use (2) and (3) for copying and readback practice. (See also CLEARANCE SHORTHAND, page 77).

POSITION REPORTS also follow a set pattern. This information is transmitted in the following order:

Station
Callsign
Reporting Point
Time
Flight Level
Next Reporting Point
ETA at next reporting point, as in the example opposite.

Mandatory postion reports are made at those positions marked by a "blocked" triangle:

whilst those marked with an "open" triangle:

are for "on-request" position reports only.

AIRWAYS CLEARANCES	
YOU	**AIR TRAFFIC CONTROL**
	(1) London Control clears G-RSTU to Ronaldsway via W39, A25 and R3 joining airways at GLO, at FL180, reporting RADNO and WAL, to squawk 4408 and to contact London Control on 131·2 when cleared.

(Student reads clearance back word-for-word and in the correct order).

	(2) Scottish Control clears G-STUV to Newcastle Teeside via A1 and DW11, joining airways at NGY, at FL130, reporting DCS, to squawk 4545 and to contact Scottish Control on 124·9 when cleared.

(Student readsback).

	(3) London Control clears G-VWXY to Guernsey via W17 and R1, joining airways at Midhurst at FL60, reporting HAZEL, SAMTN and BENDY, to squawk 4541 and to contact London Control on 134·45 when cleared.

(Student reads back).

The above examples are solely for practice in using that format, as reporting points would differ and transponder codes vary according to airway and region.

Lesson 18: Airways' Procedures

— is a simulated flight test on Instrument Rating Test Route SSIR 30 from Birmingham International to Stansted. You can follow this flight on AERAD EUR 1&2 or Jeppeson E(Lo) 1&2.

This lesson is intended to demonstrate airways procedure to the FRTO holder who wishes to gain an instrument rating and so fly on airways. It includes the following calls that are part only of airways procedure:

airways clearances
airways joining
airways position reports
leaving airways

It also demonstrates a holding en-route call, ground movement control and how to check both radios on more than one frequency, as in an actual IR test. The sequence for this is:

VHF Box	Station to call	Request
2	Tower	"Radio Check."
	(Set and check altimeters)	
2	Approach	"Radio Check Box 2."
1	Approach	"Radio Check Box 1."
1	Tower or Ground	"Request taxi clearance."

This means that Box 1 is the radio in use and Box 2 is ready for use.

This "test" is conducted in a Pa34 Seneca. The callsign "TEST 4" is a designator for an aircraft on a CAA test.

Radio traffic begins at (1) with the candidate obtaining the airfield data from ATIS, in this case, Information BRAVO.

(2) begins the sequence for checking both radios on more than one frequency and ends up with the call to ground movement control for taxi clearance (3).

The candidate wisely chooses a convenient time to copy his airways clearance (4) & (5), reads it back verbatim, and changes to the tower frequency for take-off clearance (6).

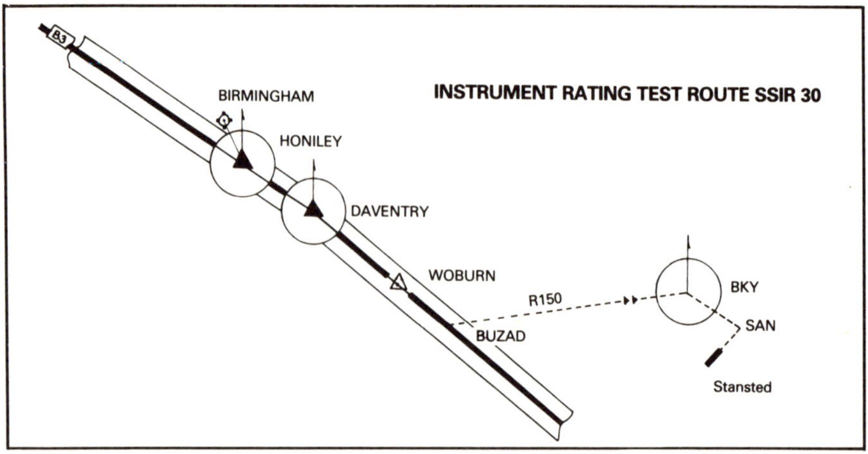

AIRWAYS	
YOU	**AIR TRAFFIC CONTROL**
(1) TEST 4 is on the apron, having gained airfield information BRAVO from ATIS on 120·7.	
(2) Birmingham Tower, TEST 4 with BRAVO. Radio check Box 2 on 118·3.	TEST 4, Birmingham Tower reading you fives. QNH1020.
Reading you fives also. QNH1020, TEST 4 changing to Approach on 120·5.	TEST 4, that is correct.

YOU	AIR TRAFFIC CONTROL
Birmingham Approach, TEST 4, radio check Box 2 on 120·5.	TEST 4, Birmingham Approach reading you fives.
Reading you fives also. TEST 4 changing to Box 1.	

YOU	AIR TRAFFIC CONTROL
Birmingham Approach, TEST 4, radio check on Box 1.	TEST 4, Birmingham Approach reading you fives on Box 1.
Reading you fives also. TEST 4 is changing to Birmingham Ground on 121.8.	TEST 4, roger.

YOU	AIR TRAFFIC CONTROL
(3) Birmingham Ground, TEST 4 requests taxi clearance.	TEST 4 is cleared to taxi to Holding Point Alpha for runway 15.
"Cleared to taxi to Hold Alpha for Runway 15," TEST 4.	TEST 4, that is correct.

YOU	AIR TRAFFIC CONTROL
	TEST 4, are you ready to copy your airways clearance?
(4) TEST 4, standby.	

YOU	AIR TRAFFIC CONTROL
TEST 4, ready to copy.	(5) TEST 4, your airways clearance is as follows: London Control clears TEST 4 to Stansted via B3, joining airways at HONiley, level at FL70, reporting DavenTrY, WOBURN and BUZAD, to squawk 4541, and to contact London Control, when cleared, on 133·7.
"London Control clears TEST 4 to Stansted via B3, joining airways at HONiley, level at FL70, reporting DavenTrY, WOBURN and BUZAD, to squawk 4541, and to contact London Control, when cleared, on 133·7."	TEST 4, that is correct. Will you be ready for departure upon reaching?

Affirm, TEST 4.	Test 4, change now to Birmingham Tower on 118·3.
"Birmingham Tower 118·3, Test 4."	

(6) Birmingham Tower, Test 4 is ready for departure.	Roger. TEST 4 is cleared for take-off, turning on track, initially not above 2000′, wind 180 less than 5 kts.
"Cleared for take-off, turning on track, not above 2000′," TEST 4.	TEST 4, that is correct.

	TEST 4 was airborne 05. Contact Approach on 120·5.
Changing to Approach on 120·5, TEST 4.	

(7) Birmingham Approach, this is TEST 4 on 120·5.	TEST 4, Birmingham Approach. No height restriction now. You are cleared to climb en-route. Call Honiley.
"Call Honiley," TEST 4.	TEST 4, that is correct.

TEST 4 is Honiley, passing 3000′, for FL70.	TEST 4, roger. Call now, London Control on 133·7.
Changing to London Control on 133·7, TEST 4.	

(8) London Control, TEST 4, Request clearance to enter controlled airspace north-west of Daventry at FL70.	TEST 4, enter controlled air-space north-west of Daventry level at FL70 cleared to DTY via B3 to maintain FL70.
"TEST 4 is cleared to enter controlled airspace north-west of Daventry level at FL70 cleared to DTY via B3 maintain FL70."	TEST 4, that is correct. Report Daventry.
"Report Daventry," TEST 4.	

(9) London Control, TEST 4. Daventry 18 level at FL70 WOBURN 26.	(10) TEST 4, roger. Descend immediately FL60 due traffic. Report level.
(11) TEST 4 leaving FL70 immediately for FL60. Will report level.	TEST 4, roger.

(12) TEST 4 level at FL60.	TEST 4, roger. Report WOBURN FL60.
"Report WOBURN FL60," TEST 4.	TEST 4, that is correct.

(13) London Control, TEST 4 is WOBURN 25, FL60 BUZAD 31.	TEST 4, roger. Report BUZAD FL60.
"Report BUZAD FL60," TEST 4.	TEST 4, that is correct.

(14)	TEST 4 hold at BUZAD FL60, expect onward clearance at 36, holding delays at BKY.
(15) Hold at BUZAD FL60, TEST 4, expect onward clearance at time 36.	TEST 4, correct.

(16)	TEST 4 is cleared to leave controlled airspace east of BUZAD, cleared to descend to 3000′ on the Chatham QNH of 1009. Call Luton Approach on 129·55.
"Cleared to leave controlled airspace east of BUZAD, cleared to descend to 3000′ on the Chatham QNH of 1009, call Luton Approach on 129·55," TEST 4.	TEST 4, correct.

(7) is a standard frequency change to Approach, and (8) is the abbreviated request for clearance to join airways as planned.

(9) is an airways position report.

In (10), the word "immediately" denotes exceptional circumstances indicating immediate action. The words "immediately" or "now" must be included in reading back required information (11).

(12) is a standard "level" call.

(13) is an airways position report.

In (14), the candidate is surprisingly requested to hold at a reporting point and (15) is his acknowledgement.

In (16), he is cleared to leave airways.

Further Study

Study pp 85-90 CAP413,
AERAD London Area chart,
AERAD EUR 1&2,
Jeppesen E(Lo) 1&2.

(That concludes the AIRWAYS section of instrument rating test route SSIR 30.)

Memory Aids

Lesson 1: INBOUND CALL
Station
Callsign
Type
(P) Position
(H) Heading
(A) Altitude/Flight Level
(C) flight Conditions
(E) ETA at zone boundary
(R) "Request Joining Instructions."

Lesson 8: VDF
INITIAL CONTACT CALL
Station
Callsign
Request (QDM or QDR)
Callsign

(ATC: "Pass your information")

QDM/QDR readback
Class
Callsign
Change of frequency if applicable

Lesson 12: DISTRESS
"MAYDAY, MAYDAY, MAYDAY."
Station
Callsign
Type
(N) Nature
(I) Intention
(P) Position
(A) Altitude/Flight Level
(H) Heading
(Q) Qualification, if desired

Lesson 13: URGENCY
"PAN, PAN, PAN! PAN, PAN, PAN!"
Station
Callsign
Type
(N) Nature
(I) Intention
(P) Position
(A) Altitude/Flight Level
(H) Heading
(Q) Qualification of pilot

Lesson 7: MATZ PENETRATION
Station
Callsign
Type
"Request MATZ Penetration"
(DETAILS)
Callsign
(P) Position
(H) Heading
(A) Altitude/FL
(C) Conditions
(E) ETA at MATZ boundary

Lesson 11: LARS
INITIAL CONTACT CALL
Station
Callsign
"Request Lower Airspace
Radar Service"

(ATC: "Pass your information")

Callsign
Type
(P) Position
(H) Heading
(A) Altitude
(C) Conditions
Any other information
e.g. transponder code

CANCEL DISTRESS
"MAYDAY" Station
Callsign
"Cancel Distress"
Reason for cancellation
Intention e.g. diversion

Lesson 14: PRACTICE PAN
"PRACTICE PAN, PRACTICE PAN,
PRACTICE PAN"
Station
Callsign
"REQUESTING PRACTICE PAN"

When instructed, go ahead with
"PRACTICE PAN, PRACTICE PAN,
PRACTICE PAN"
Station
Callsign
Type
(N) Nature
(I) Intention
(P) Position
(A) Altitude
(H) Heading
(Q) Qualification

Lesson 16: AIRWAYS CROSSING
Station
Callsign
(R) "Request Crossing Clearance" at (Crossing Point)
DETAILS
Callsign
Type
(P) Position
(H) Heading
(A) Altitude
(H) Heading
(E) ETA and desired FL at Crossing Point

Lesson 17: POSITION REPORT
Station
Callsign
Position and
Time at the position
FL
Next reporting point
ETA at that reporting point

Suggested Clearance Shorthand

The following is by no means an exhaustive list of timesavers for the more commonly-heard clearance instructions. They are personal shortcuts tried and test and added to from time to time. Try them if you find they save you time. Add or modify any you wish.

QFE	E	QNH	H
Line up	LU	Hold	Hd
Cleared for take-off	CTO	Call before	Call>
Left turn out	LT	Right turn out	RTO
Not above (altitude)	⊤	Not below	⊥
Climb and maintain	⟋‾→	Descend and maintain	⟍_
Out of control zone	⊖→	Through control zone	-⊖→
Before	>	After	<
Cross	X	At	@
Cross airways	╪	Until	/
Join airways	═→	While on airways	=
While in zone	○	Transponder	TX
Enter control zone	→⊖	Entry lane	EL
Report at	R@	Report passing	RP
Report leaving	RL	Report reaching	RR
Clearance void after	CV<	Special VFR	SVFR
Expect further clearance	EFC	Contact (name, freq.)	124·6

Glossary of Terms

"ACKNOWLEDGE": see p.11.

ACTUAL: A report of the weather ACTUALLY occurring, as distinct from that forecast.

"AFFIRM": see p.11.

AICs: Aeronautical Information Circulars.

AIP: The United Kingdom Aeronautical Information Pilot, or UKAIP.

AIRWAYS: an area of controlled airspace in the form of a corridor, the centre of which is determined by a navigational beacon.

ANO: Air Navigation Order.

"APPROVED": see p.11.

ATC: Air Traffic Control — in this publication, meant to include all personnel operating ground radio stations.

ATIS: Automatic Terminal Information Service, a continually-broadcast tape-recording of routine information for arriving and departing traffic, e.g. runway in use, pressure settings.

ATS: Air Traffic Service — the provision of flight information, alerting, air traffic advice, air traffic control, approach control or aerodrome control services.

ATSU: Air Traffic Services Unit — the ground personnel providing any of the above services.

AUDIO PANEL: where more than one radio is fitted, the audio panel enables the pilot to select radio Box 1 or Box 2, and to choose whether to listen to it via headphones or via the cabin speaker. Usually mounted at the top of the radio stack.

AVIONICS: an amalgam of the words "AVIation electrONICS."

"BLIND TRANSMISSION": The phrase "transmitting blind" alerts others to the fact that the person transmitting suspects transmitter or receiver failure.

BOX: as in "radio box", means a "radio set", or apparatus.

"BREAK": see p.11.

BROADCAST: Transmission intended for general reception and not for a specific station.

"CANCEL": see p.11.

CAP413: Civil Aviation Publication no.413, a general reference published by the CAA.

CHANNEL-SPLITTING: the further narrowing of frequency spacing to allow for an increase in the number of aeronautical VHF channels.

"CHECK": see p.11.

"CLEARED": see p.11.

"CONFIRM": see p.11.

"CONTACT": see p.11.

"CORRECT": see p.11.

DF (VDF): Directing-Finding.

"DISREGARD": see p.11.

EAT (Expected Approach Time): the time at which an aircraft has been instructed to begin its approach to land, given in case of a subsequent failure in communications.

EPTT: Emergency-press-to-transmit: generally a toggle switch mounted near a radio stack, used when the operation of the normal transmit button is suspect.

FRTO: Flight Radio Telephony Operator usually known as R/T licence or RTF licence.

GMT: Greenwich Mean Time. For all practical purposes meaning the same as Universal Co-ordinated Time. Often referred to as Zulu Time or "Z".

"GO-AROUND": see pp.22 and 23.

HEADSET: a combination of earphones and microphone.

HOMER: spoken after the callsign of an aerodrome, this indicates to the personnel that the speaker will require a VDF service.

"HOW DO YOU READ?": see p.11.

IF: flight according to Instrument Flight Rules.

IFR: Instrument Flight Rules.

INTERCOM: apparatus enabling communication via headsets between pilots, or pilots and passengers, generally fixed but can also be portable.

"I SAY AGAIN": see p.11.

kHz: thousands of times per second, replaced "kilocycles". See also mHz.

LARS: Lower Airspace Radar Service.

"LATE DOWNWIND": see p.21.

LATCC: (pronounced "Latki") London Air Traffic Control Centre.

"LINE OF SIGHT": travelling in a straight line and not following the curvature of the earth.

LED: light-emitting diode — a type of frequency display.

"LONG FINAL": between four and eight miles when on a straight-in approach and lined-up on the extended centre-line of the runway.

MATZ: Military Aerodrome Traffic Zone.

mHz: millions of times per second. See also kHz.

MODE: "Mode charlie" transponders provide the radar operator with the altitude information of an aircraft as well as its "squawk" number, and "mode alpha" transponders do not.

"MONITOR": see p.11.

"NEGATIVE": see p.11.

NIPAHQ: see memory aids, p.76.

"ORBIT": this means to fly a continuous 360-degree turn at rate one, thus delaying your aircraft for two minutes to allow for traffic separation, generally in the circuit.

"PASS YOUR MESSAGE": see p.12.

PHACER: see memory aids, p.76.

PROGRESSIVE TAXI INSTRUCTIONS: to be asked for if uncertain of layout of taxiways.

"READBACK" : see p.12.

"REPORT": see p.12.

"REQUEST": see p.12.

"ROGER": see p.12.

RTF: radiotelephony, or more commonly, R/T.

"RUNWAY VACATED": see p.23.

"SAY AGAIN": see p.12.

SIDETONE: is the phenomenon whereby the pilot can hear his/her own voice when transmitting, and is a good indication that the microphone is functioning properly.

SQUAWK: means set your transponder to "Standby". Then dial in the following number and turn the transponder to "on", — as in: "Squawk 1771".

SQUELCH: is the unwanted "noise" on a received message. By increasing sensitivity to squelch it is sometimes possible to understand a message which would otherwise be unreadable.

SSR: Secondary Surveillance Radar; where a transponder in the aircraft is triggered to send a numbered recognition signal back to the radar display, as distinct from Primary Radar, which merely shows a reflected signal in the same way that rain clutter or a flock of birds does.

"STANDBY": see p.12.

"TRANSMIT BLIND": see "Blind Transmission".

UCT: see GMT.

VDF: Very High Frequency Direction-Finding, more commonly DF.

VOR: Very high frequency Omni-directional Radio range.

VERBATIM: in this context, it means word-for-word, and in the same order.

"VERIFY": see p.12.

"WILCO": see p.12.

XMTR: shorthand for "transmitter".

ZULU or Z: see GMT.

INDEX

References to illustrations in **BOLD**.